Asterisk® For Dummies®

Cheat Sheet

Installation and Configuration

Files you need during installation are as follows:

- Asterisk file named `asterisk-1*.tar.gz`
- Asterisk sounds named `asterisk-sounds-*.tar.gz`
- Zaptel file named `zaptel.-*tar.gz`
- Libpri file named `libpri-*tar.gz`

Files you need to configure your Asterisk are as follows:

- `Zaptel.conf`: This connects your Zaptel cards to the Linux kernel for analog transmissions.
- `Zapata.conf`: This identifies the Zaptel cards that Asterisk uses.
- `SIP.conf`: You can set VoIP parameters in this file for SIP protocol.
- `IAX.conf`: This handles the finer details of the VoIP protocol used for InterAsterisk eXchange calls.

Analog configurations

After installing Zaptel analog cards to your server, configure each port on the card based upon whether you have an FXO or FXS port. Don't forget that the card's design as an FXS or FXO port is opposite the type of signaling you're programming it for:

- FXO module with FXS signaling is required to interface Asterisk to your carrier.
- FXS module with FXO signaling is required to interface Asterisk to a phone in your office.

VoIP formatting

The specific SIP parameters are established in the `SIP.conf` file. Each context within that file falls in one of three categories:

- **Friend:** This classification allows the context to send and receive VoIP calls.
- **Peer:** Identifying a peer allows the device associated with the context to only make outbound calls, but it cannot receive any VoIP calls.
- **User:** The user classification is the opposite of the peer classification; it's only allowed to receive incoming VoIP calls.

Troubleshooting your code

Asterisk allows you to debug your software from the Asterisk console with a `debug` command in the following format:

`modulename debug <parameters>`

For example, you use this command to troubleshoot a VoIP port using SIP on IP address 207.111.170.18:

`Sip debug IP 207.111.170.18`

For Dummies: Bestselling Book Series for Beginners

Asterisk® For Dummies®

Cheat Sheet

Dialplan Basics

You configure the dialplan in the `extensions.conf` file. The code appears similar to the following, with four constituent parts:

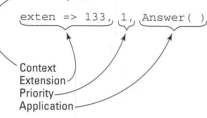

```
[incoming]
exten => 133, 1, Answer( )
```

Context
Extension
Priority
Application

- **Context:** The context is always presented in the dialplan surrounded by brackets and groups all the code beneath it.
- **Extension:** The extension, in this example 133, is a destination for a call. It may be a physical phone or an internal feature designated to receive the call.
- **Priority:** Priority creates a numerical sequence of events that Asterisk follows.
- **Application:** Application is a logic-based decision or action applied to a call. The application may require `DIAL`, `GOTO`, or `GOTOIF`, or even access a macro. It usually contains the logic for the decision-making process of the application. Some applications are as follows:
  - `Set(variable=value)`
  - `Goto([context,]extension,priority)`
  - `MeetMe([[confnumber] [,[options] [pin]])`
  - `Queue(Queue_Name[,Options[,URL[,Announce_Override[,Time_Out]]]])`
  - `GotoIf(condition?label1:label2)`: You can have a priority, an extension and a priority, or a context, extension and a priority as labels.

Pattern matching

Pattern matching begins with an underscore (_) in all code, with variables representing the following numbers:

- `_N`: Numbers 2–9
- `_X`: Numbers 0–9
- `_Z`: Numbers 1–9
- `_.`: All whole numbers

Voice-mail setup

```
vmexten => password,name,e-mail,pager,options
```
The options in the voice-mail extension code are separated by pipes (|).

Dial-by-name directory

Dialing by first name uses the following code:
```
directory (default,incoming,f)
```

Dialing by last name uses the following code:
```
directory (default,incoming,)
```

For Dummies: Bestselling Book Series for Beginners

Asterisk®

FOR

DUMMIES®

Asterisk® FOR DUMMIES®

by Stephen P. Olejniczak and Brady Kirby

Foreword by Mark Spencer
Creator of Asterisk

BICENTENNIAL
1807
WILEY
2007
BICENTENNIAL

Wiley Publishing, Inc.

Asterisk® For Dummies®

Published by
Wiley Publishing, Inc.
111 River Street
Hoboken, NJ 07030-5774

www.wiley.com

Copyright © 2007 by Wiley Publishing, Inc., Indianapolis, Indiana

Published by Wiley Publishing, Inc., Indianapolis, Indiana

Published simultaneously in Canada

For general information on our other products and services, please contact our Customer Care Department within the U.S. at 800-762-2974, outside the U.S. at 317-572-3993, or fax 317-572-4002.

For technical support, please visit www.wiley.com/techsupport.

Wiley also publishes its books in a variety of electronic formats. Some content that appears in print may not be available in electronic books.

Library of Congress Control Number: 2006939592

ISBN: 978-0-470-09854-7

Manufactured in the United States of America

10 9 8 7 6 5 4 3 2 1

WILEY

About the Authors

Stephen Olejniczak (pronounced *ō' lĕn ēē chek*) is the Director of Operations for ATI, and has over 14 years of experience in telecommunications. His experience is focused on installation, service, and support of voice service with a growing responsibility for delivering VoIP. He's the author of *Telecom For Dummies* and a snappy dresser.

Stephen didn't start out in life as a techie, only falling prey to the glamour and easy money after failing to find a career enabling him to use his Bachelors degree in Cultural Anthropology. He currently lives in the quaint hamlet of Laguna Beach, California with his beautiful wife Kayley and a collection of fountain pens.

Brady Kirby (pronounced *Kir'bēē*) is the owner of Atlas VoIP and has been programming since he was 13 years old. The initial challenge of writing small screen savers in qbasic wasn't enough for him and by the 10th grade he worked his way up to the strong stuff — game programming. It was at this tender age that Brady accepted his calling. He realized it was his lot in life to make the world a better place, solving people's problems, by evolving inventive software.

Brady spends his free time with his son, Brady Kirby III, in the crucible of technology and engineering, Huntsville, Alabama. Surrounded by many tech companies, and that space program we have (we think it's called NASA), they enjoy the good life with large helpings of lasagna and good Mexican food. It's in this environment that Brady III is mentored in all aspects of programming by his father. Maybe to write his own *For Dummies* book someday.

Dedication

We dedicate this book to the brave men and women, writing programs and evolving code, and allowing globally available open-source software to flourish. As beneficiaries of your hard work, we take this time to say "thanks" to those people who like to stay up until 3:00 AM writing programs and eating cold pizza or quickly heated Raman noodles. We also acknowledge all of the new entrants into this area and hope this book provides a foundation and guidance for your initiation into this community. And finally, we'd like to pay our respects to the early anthropolithicines that were fascinated with tools, craved sugar and salt, and were amorous enough to make the lasagna-consuming code-typing programmers of today possible.

Authors' Acknowledgments

Stephen P. Olejniczak: I must first thank my beautiful wife Kayley for giving me the time necessary to write this book. Her grace is only matched by her intelligence and kindness and I am the luckiest man in the world to be married to her. I also thank my co-author, Brady Kirby, who is the intellectual powerhouse of this book. In spite of the fact that I might not have initially understood what you were talking about, eventually we got it all broken down so regular folks like me could understand. Along the same lines I thank my agent Matt Wagner who brought this project to me, our acquisition editor Melody Lane who was always professional and supportive, and our editor Rebecca Senninger who exhibited the patience of a saint when we hit some rough spots. Finally, I thank my late mother Diane Olejniczak who inspired me to write my first book. Her biography, edited and published by my sister Christine Olejniczak Freeze, was my first serious attempt at writing. The process taught me a lot and paved the way for *Telecom For Dummies*, as well as this book.

Brady Kirby: I'd like to thank my son, the greatest gift a man can receive, for being a wonderful child, and for always keeping my motivation for life going, even during the writing of this book. Without the support of Tim Erwin of Traveller Internet Solutions and Atlas VoIP Communications for giving me the opportunity to create this new company, learn more about this great software, Asterisk, and also James Staley and Stephen Olejniczak (my co-author) of ATI for the opportunity to learn more about the telecom side of voice, and in writing this book we would have not made it this far. I also have to say thanks to all the customers and co-workers who have ever asked a "why does it do this..." or "can we make it work like this..." question over the past couple of years, and have let me pursue not only learning the software, but also in creating a self-sufficient company for supplying VoIP services. Finally I'd like to thank everyone at Wiley Publishing who have made *Asterisk For Dummies* a possibility; without you none of this would be possible.

Publisher's Acknowledgments

We're proud of this book; please send us your comments through our online registration form located at `www.dummies.com/register/`.

Some of the people who helped bring this book to market include the following:

Acquisitions, Editorial, and Media Development

Project Editor: Rebecca Senninger

Acquisitions Editor: Melody Layne

Copy Editor: John Edwards

Technical Editor: Carl Doss

Editorial Manager: Leah Cameron

Media Development Manager: Laura VanWinkle

Editorial Assistant: Amanda Foxworth

Sr. Editorial Assistant: Cherie Case

Cartoons: Rich Tennant (`www.the5thwave.com`)

Composition Services

Project Coordinator: Jennifer Theriot

Layout and Graphics: Carl Byers, Brooke Graczyk, Stephanie D. Jumper, Laura Pence

Proofreaders: John Greenough, Charles Spencer, Techbooks

Indexer: Techbooks

Anniversary Logo Design: Richard J. Pacifico

Publishing and Editorial for Technology Dummies

   Richard Swadley, Vice President and Executive Group Publisher

   Andy Cummings, Vice President and Publisher

   Mary Bednarek, Executive Acquisitions Director

   Mary C. Corder, Editorial Director

Publishing for Consumer Dummies

   Diane Graves Steele, Vice President and Publisher

   Joyce Pepple, Acquisitions Director

Composition Services

   Gerry Fahey, Vice President of Production Services

   Debbie Stailey, Director of Composition Services

Contents at a Glance

Table of Contents

Foreword

*W*ay back when I was but a wee lad (okay, like 9 years ago as a freshman in college), and trying hard to convert all my friends over to Linux, I remember reading a post from someone who said, "If Linux is so great, how come there's no *Linux For Dummies* book," to which someone replied "That's because dummies don't use Linux." Of course for quite a while now there has been a *Linux For Dummies* book and slowly, more and more people have been able to learn and take advantage of that technology because of it. Similarly, writing a foreword for *Asterisk For Dummies* marks an important mile marker in Asterisk's growth and transition from a niche technology to a core technology.

Many people think that Asterisk started as a grand vision for what telephony would one day become. In reality, it started out from a much more humble and pragmatic reason: I started a company, Linux Support Services, with only a few thousand dollars of startup capital, needed a phone system and they were just too expensive. So as I built my own computers, did my own Web site, made my own work benches and so on, I figured I'd write my own phone system as well. Having been involved in open source (free software) for some time (for example I was the original creator of gaim, the popular instant messenger), it was only natural for me to make my creation available free of charge, right from the start. And for better or worse, I rarely think small, so I chose the name "Asterisk" from the wildcard symbol of Linux, UNIX, and DOS ("*") meaning "everything".

A couple of years later, I renamed the company Digium and began focusing on Asterisk as our core business (which allowed me to continue to fuel Asterisk's advancement). The telecom industry has been explosively ready for open source and Asterisk has been well received with over 400 contributing developers and at the time of writing, at least 130 companies with their businesses built around Asterisk. So what's the big deal that makes Asterisk so powerful? How many times have you thought to yourself, 'You know, it'd be really cool if my phone system could do...'? Well, Asterisk allows people to turn ideas into products and services remarkably quickly (including alarm systems, paging gateways, systems to let you pay for parking spaces, systems for making international cell phone calls free, and more). This book, along with advances in software and hardware appliances, is bringing Asterisk out of being the best kept secret in telephony and into the mainstream. I'd buy some Red Bull before you get started though; it's going to be a long night!

– Mark Spencer, creator of Asterisk

Introduction

. . . .

A strong division has always existed between the people in telecom who work the data applications — such as Multiprotocol Label Switching (MPLS), Frame Relay, and Internet circuits — and the people who work with hardware that only passes voice phone calls. The voice technicians devote their time to understanding a completely different realm of information than their data counterparts. For some reason, the split has always been there and has been reinforced by corporate structuring, sales, and support. Asterisk changes that.

Asterisk is truly a philosophical bridge in the world of telecom. It takes the methodical programming and code-driven design aspects of the *data* side of telecom and directly applies them to the handling and processing of *voice* calls. The mere prospect of such a system is tremendous, but the challenge is that few people have experience in both the voice and the data aspects of telecom.

The new frontier of telecom ushered in by Asterisk demands a person with skills on both sides of the fence to make it work. We wrote this book because this type of person doesn't normally exist in our industry. The readers with a programming background find everything they need to install, program, and grow with Asterisk. The readers without a background in programming also find out about basic Linux, Voice over Internet Protocol (VoIP), and programming.

About This Book

Asterisk For Dummies is unlike any other Asterisk book on the market. It's designed for everyone with a desire to use Asterisk in a home or office, or as the basis of a business. We realize that Asterisk is one piece of your entire telecom infrastructure. It doesn't just exist independent of the world; it interacts with your local-area network (LAN) and your local and long-distance carriers. As such, we address the entire environment in which Asterisk lives, not just as antiseptic software living in a lab. We have used Asterisk as the basis for a carrier platform for several years, and we give you all the wisdom we have gained over that time.

Conventions Used in This Book

As you go through the book, you'll notice we use special formatting for certain things. There's a method to our madness; here's a summary of the formatting we use:

- ✔ Words in *italic* indicate new terms that we define for you.
- ✔ Code, filenames, commands, Web addresses, and on-screen messages appear in monofont.
- ✔ **Bolded monofont** highlights a particular section of the code we're discussing.
- ✔ *Italic monofont* indicates a placeholder; you should replace the placeholder with your specific information. For example, when you see this

```
grep mysearchterm *
```

  you replace *mysearchterm* with your specific search term.
- ✔ All dialplan contexts are surrounded by brackets, such as [incoming].

What You Don't Have to Read

As a standard rule, don't read what you already know. The fact that Asterisk brings together both computer-programming people and voice telecom people means that each group has its own strengths and weaknesses. If you're a telecom person, you may want to skip the basic troubleshooting chapters. If you're a programming person, you may want to just skim the dialplan chapters. Don't worry, we won't take offense.

You're also free to skip the text that's next to the Technical Stuff icons — unless, of course, you like finding out the geeky info that these icons highlight.

Foolish Assumptions

We begin by assuming that you're interested in Asterisk and have a technical background of some sort. You may have been working as a programmer, writing code for the past few years, or as a dedicated provisioner, tuning up T-1 circuits for a long-distance carrier. In any case, you're a techie who uses computers on a daily basis.

The more important thing is what we aren't assuming. We don't assume that you have knowledge of Linux or programming. We don't assume that you know what a root directory is, how to modify a document in Linux, or how to create a tarball.

We do recommend knowing as much as possible about Linux and programming through any of the *For Dummies* books, such as *Red Hat Fedora Linux 2 For Dummies,* by Jon Hall and Paul G. Sery, and *C++ For Dummies, 5th Edition,* by Stephen Randy Davis (both published by Wiley). Knowing about Linux operating software and knowing how to write computer programs make working with Asterisk easier.

If you're a programmer who is installing Asterisk, we recommend *Telecom For Dummies,* by Stephen P. Olejniczak (published by Wiley) as a basic primer on the telecom industry. It covers all the aspects of telecom ordering and provisioning and is a wealth of information on troubleshooting all varieties of problems.

How This Book Is Organized

This book follows a normal life-cycle progression, from concept to installation and management. The chapters cover evolving levels of development, but you don't need to have read the previous chapter to understand the information presented in a given chapter.

Part 1: Introducing . . . ASTERISK!

This part begins with an introduction to Asterisk, its capabilities, and its potential. Then, we identify where to acquire and install Asterisk software and the AsteriskNOW software. Finally, we discuss the hardware interfaces for Internet, analog, and digital connections.

Part 11: Using Dialplans — the Building Blocks of Asterisk

In Part II, we focus on the Asterisk dialplan. We show you how to construct routes within Asterisk to handle inbound and outbound calls, and we describe voice mailboxes, call queues, and conference rooms. If you're using the AsteriskNOW software, you find out how to configure it. We finish off the part by discussing the Asterisk database and VoIP codecs.

Part III: Maintaining Your Phone Service with Asterisk

This part covers everything you need to know to keep your Asterisk system functioning. We provide you with the tools necessary to troubleshoot the calls handled by Asterisk. We describe how to capture VoIP information, identifying the source of transmission issues. We also cover general and specific troubleshooting of analog, digital, and InterAsterisk eXchange (IAX) transmissions.

Finally, we cover Asterisk call volume management and discuss hardware and server maintenance requirements, including disaster recovery and security.

Part IV: The Part of Tens

In The Part of Tens, we cover ten things you shouldn't do with your Asterisk, ten fun things to do with your Asterisk, and ten places to go for help.

Part V: Appendixes

You just may find the three appendixes vital to your overall success with Asterisk. Appendix A provides an overview of the Asterisk dialplan you must write to route calls with Asterisk. This first appendix brings together all the specific dialplan functions from Part II, enabling you to see the connections.

Appendixes B and C cover VoIP and Linux. We provide a basic primer on each, with additional information geared toward the specific skills required to install, manage, and troubleshoot Asterisk.

Icons Used in This Book

This book includes nifty icons in the margins. They indicate info that you'll find helpful.

The Tip icon points out shortcuts that can save you some time and effort. If you like taking the shortest route, you'll like these tips.

The Remember icon highlights info that we think is absolutely, undeniably important to remember. Don't forget it.

 Be sure to pay attention to the Warning icons. They may just save you from potential trouble.

 The Technical Stuff icon highlights info that only the truly geeky find interesting.

Where to Go from Here

Start with the Table of Contents. Where you go from there is entirely up to you. If you're working on an Asterisk system that's already installed, the dialplan basics in Chapter 5 might be the first thing you read. If you have a VoIP problem that's been plaguing you, read Chapter 10, which discusses Wireshark. Every chapter is independent, and you don't have to read every previous chapter. If you need more information, we direct you to the chapter that explains it in detail. So scan the Table of Contents, see what interests you, and enjoy!

Part I
Introducing . . .
ASTERISK!

The 5th Wave   By Rich Tennant

"I can be reached at home on my cell phone, I can be reached on the road with my pager and PDA. Soon I'll be reachable on a plane with e-mail. I'm beginning to think identity theft wouldn't be such a bad idea for a while."

In this part . . .

We introduce you to Asterisk in all its glory. If you're just starting out with Asterisk, dive into this part first. Chapter 1 covers the basic functionality of Asterisk. We cover the hardware requirements and discuss how Asterisk can replace your current business phone system or act as the backbone of a value-added telephony business. Chapter 2 goes into the specifics of acquiring the correct version of Asterisk. We show you where to download the software and how to load it onto your Linux server. Chapter 3 shows how to download and install AsteriskNOW. We round out the part with Chapter 4, where you discover how to configure the telephony ports that access the analog, digital, InterAsterisk eXchange, and VoIP connections linking you to the world.

Chapter 1

Evaluating the Possibilities with Asterisk

*T*he sophistication of telecom hardware available to the average company has increased tremendously. It was amazing in the 1980s when you could use an automated dialer to direct calls to your long-distance carrier instead of your local carrier. In the 1990s, the technology and market had evolved to where you could identify calls based on where you were dialing and match them to the carriers you may have had at your disposal for the best rate. In 2000, all the buzz was to use Internet circuits to transmit calls. At the time, the quality was a bit suspect, but most of those issues have been worked out, and people now send Voice over Internet Protocol (VoIP) calls around the world every day.

This evolution created the environment that spawned Asterisk. A division in telecom has always existed between the engineers that transfer data and the engineers that work on voice service. In general, these are different types of people, and they don't hang around with each other. Asterisk brings them both together for the first time. It allows you to handle voice calls with the same mind-set of routing data.

This chapter introduces you to the depth and breadth of what you can do with Asterisk. We cover the general types of calls and features as well as the varieties of telephony platforms you can use to interact with Asterisk. We also cover the hardware necessary to make and receive calls, and then we give you the parameters for the server that you need to run your Asterisk for optimum performance.

Finding Out What You Can Do with Asterisk

Asterisk bridges the methodical, planned open-source architecture previously the sole domain of data transmissions and takes it into the voice world dominated by rigid proprietary phone systems. The intelligence and flexibility of telecom hardware available to the person on the street has grown at a very fast pace over the past 20 years.

In the future, Asterisk has the power of bringing together people from around the world for free. VoIP connections handled entirely across the public IP network don't incur per-minute charges like a normal phone call. Only when the proprietary networks of the long-distance and local phone companies handle a call do per-minute charges apply. Otherwise, the transmission travels across your office or across the world for free, just like an e-mail.

In the sections that follow, we show you what you can do with your Asterisk.

Using Asterisk for your phone system

In the beginning, Asterisk was rumored to have been designed as a voice-mail system. That is by no means the limit of its potential; voice mail is simply the beginning of its capabilities. The software now functions as a platform for receiving and transferring calls with all the standard features you want from a phone system, such as

- Voice mail, which allows callers to record a voice message when you are on your phone line or away from the office
- Conference calling, which provides the ability for multiple people to call into your Asterisk and talk together just as if they were all physically sitting in the same conference room
- Dial-by-name directory, which allows callers to reach an extension by spelling out the first or last name of the person of the person they are calling on the keypad of their phone
- Call parking, which sends a call someplace to wait on hold
- Music on hold, which gives callers something to listen to while you have them on hold

These same features are available with any phone system. You need to purchase hardware to allow you to use either dedicated digital (T-1) lines or regular, analog, plain old telephone service (POTS) lines.

You can use Asterisk as your phone system without ordering any standard analog or digital phone lines from your local carrier. As long as you have a dedicated Internet connection to your home or office, you can purchase incoming and outgoing phone service from your Internet Service Provider (ISP) and never have to worry about standard phone lines.

The main benefit of Asterisk is its flexibility. You aren't locked in to any preset parameters of the phone system. Some of the great features you have complete control over with Asterisk are as follows:

✔ How you receive calls

✔ Whether a call is sent directly to voice mail

✔ Whether a call is sent directly to a conferencing room, where several people can speak on the same conversation

✔ How calls are routed within your phone system and your company

✔ How to forward calls (such as first to your landline, then to your cellphone, and then to voice mail)

✔ How much time you allot before you rescind the call and transfer it to the next destination (such as a cellphone or voice mail)

✔ The extensions and features available to every phone and piece of hardware

Jumping into VoIP with Asterisk

One of the best aspects of Asterisk is the ease with which you can integrate the ability to send and receive calls over your dedicated Internet connection. *Voice over Internet Protocol,* more commonly referred to as *VoIP,* is the greatest revolution to hit telephony since the dial tone. All the benefits of VoIP have yet to be realized, but Asterisk comes standard with the ability to convert your voice call into VoIP packets.

One of the fastest growing applications of Asterisk is its use with VoIP. Many companies have sprung up for the sole purpose of providing advanced features to their customers via VoIP. The "find-me-follow-me" type of service that allows one call to attempt to reach you on your office line and then your cellphone, before the call is rescinded and sent to your voice mail, is just one of the great features of VoIP. You can even e-mail your voice mail as a WAV file to your BlackBerry or Palm device. Talk about never missing a call again. If you are breathing and available by any means of technology, Asterisk can find you.

The future of telecom is in the transmission of voice, fax, and video over Internet lines. The ability to packetize all these transmissions into the IP realm is a huge step forward in telecom. It allows for growth at an accelerated pace as the transmissions are handled together, maximizing the bandwidth of every connection. By breaking the mindset of traditional telephony where every voice call is locked into a 64Kbps channel, we lose the hardware requirements preventing us from using the Internet over our regular residential phone lines.

Bridging technologies of VoIP and non-VoIP

VoIP is great, but not every piece of hardware you may need to use is VoIP-based. You may use VoIP service within your office but have a better rate for out-of-state calls on a carrier that doesn't have a VoIP interface. In this case, you have to send the call with the standard, non-VoIP method using traditional time-division multiplexing (TDM) connections.

Not every long-distance carrier has the VoIP service you want. Some carriers only provide inbound VoIP service to avoid providing directory assistance, 411, and 911 services. Other carriers may avoid these requirements entirely by boycotting VoIP connections.

Your local or long-distance carrier may charge you differently for VoIP than for TDM service. The per-minute rate you are charged may be different, the area of coverage provided may be restricted, or additional monthly recurring charges or installation fees may apply.

Just because your carrier may not accept VoIP calls doesn't mean that you can't use VoIP somewhere in your system. You simply need to convert your calls from VoIP to TDM, or TDM to VoIP, when you enter or leave your carrier's network. The ability of Asterisk to act as a *gateway* (converting calls from VoIP to TDM or vice versa) gives you the freedom to use every telephony option at your disposal.

Bringing wireless into the equation

You aren't confined to traditional landlines with Asterisk, so don't get trapped by that old private branch exchange (PBX) mind-set. In the land of wireless technology, Asterisk is a good friend. It can easily interface with any of the following devices:

- ✔ Regular cordless phones
- ✔ Cordless VoIP phones
- ✔ Wireless headsets connected directly to your computer via Bluetooth

You can use most Bluetooth wireless headsets that you use with your cell-phone. You use your computer as the audio device and pair it with any VoIP software phone (*softphone*) for your PC. This configuration allows you to make calls from your softphone via your PC with VoIP. Asterisk implements its own Bluetooth channel driver, allowing Asterisk to function as your softphone on your PC.

Running your telephony business with Asterisk

Using Asterisk for your internal company calls is only the starting point of its potential. Many companies, especially VoIP-based resellers, may run their entire business on Asterisk servers. If you run a telephony business, you'll find the following useful features in Asterisk:

- ✔ Flexibility in programming
- ✔ Reporting features
- ✔ Call Detail Record (CDR) logs
- ✔ The ability to use TDM, VoIP, and wireless connections

But the fun doesn't stop there. Asterisk is easily partitioned for multiple companies to use all the available features, but still confine their calls to within their company. You don't have to worry about pressing 0 to speak to an operator at Company A and getting the receptionist for Company B. This requires a bit more effort because you must partition all the calls by company in your CDR.

Realizing the benefits of VoIP to big businesses

Multilocation corporations that have offices in different states or even countries can use Asterisk to virtually eliminate long-distance calls between offices. Installing an Asterisk at each location allows you to send and receive VoIP calls, taking your standard long-distance calls from your existing carrier. Just as e-mail eliminates the need for sending letters through the post office, VoIP now allows you to bypass your local phone carrier.

If you are using the Asterisk internally, a basic configuration should cover most of the scenarios you might encounter (see Figure 1-1, later in this chapter). If your company is a telecom reseller that provides phone service to customers, possibly as a VoIP value-added reseller, your configuration needs to be a bit different. In spite of the fact that you might have a few phones in your office attached to the Asterisk server, Asterisk primarily processes your end users' calls.

Asterisk is the best platform for VoIP resellers because you can expand and build on it as your business grows and becomes more complex. You can realize a configuration as grand as your company can become. Your setup can range from a single server that provides central call processing for multiple offices using VoIP phones to multiple servers that process calls from multiple offices that use (a) VoIP phones directly and (b) VoIP PBXs such as Asterisk to fulfill all their office needs.

A PBX is a common phone system that you would find in an office environment. It generally provides voice mail, call transfer, call hold, and the general routing of calls handled by an office phone system.

Getting Acquainted with AsteriskNOW

AsteriskNOW is a complete software package, which allows you to load Asterisk software and a distribution of Linux operating software on a server. After you complete the installation, you can configure it using an Internet GUI. All the regular features provided in Asterisk, such as voicemail, conference call setup, call queuing and dialplans, are also available through the AsteriskNOW release.

AsteriskNOW makes Asterisk available to a much broader range of companies and individuals. The idea of working in a Linux operating environment and building the routing rules with manual programming might be too much for some. For those who want to the power of Asterisk, but don't have the time to build out the programming, AsteriskNOW is your ticket.

AsteriskNOW is just like the standard version of Asterisk in its hardware requirements. It's a real-time program requiring the full and undivided attention of your server. Whether you are using the software as a phone system for your business or as a basis for the value-added telecom service you are providing, you must have it on the best hardware possible. Regardless of the size of your company or server, the Asterisk program must be given top priority in all tasks to reduce the chance of network delays degrading the quality of your calls. A modest amount of delay caused by other network activity can result in static or sections of calls being dropped. At the other end of the spectrum, large delays could result in static or the failure to connect both inbound and outbound calls.

Introducing the Supporting Hardware

Asterisk is wonderful, but it is only software diligently working inside a server. It is fully capable of converting a call from VoIP to TDM, but you still need some kind of hardware interface to connect your server to the phones in your office or the outside world. Even when you install the required hardware to make these connections, you still may need additional software drivers to bring them to life. Table 1-1 lists the hardware and software you need to make connections with Asterisk.

Table 1-1	Asterisk Interface Hardware and Drivers	
Application	*Hardware Available*	*Software Driver Required*
Analog connection	Digium FXS and FXO cards	Zapata
Digital T-1/E-1 connection	Digium T1 card for 1, 2, or 4 ports	Zapata
Digital T-1/E-1 connection	Zapata single T1 card	Zapata
Digital T-1/E-1 connection	Sangoma single T1 card	Zapata
VoIP connection	Network interface card (NIC)	ztdummy*
All above applications	No additional hardware	libpri**

The ztdummy driver acts as a timing source whenever you aren't using a POTS/T1/E1/J1 card. Applications such as conferencing and calls transferred within the Asterisk server require this timing source. The 2.6 Linux kernel is configured with an internal clock source to avoid this issue. Make sure that you download the latest driver.

**The libpri software is a required part of the Zapata modules.*

Determining your analog hardware needs

Servers don't have an unlimited supply of expansion slots available on them. Before you download your Asterisk software, you need to develop at least a two-year plan; we recommend a five-year plan. Analog cards take up the most room and are truly limiting. A four-port analog card can only handle four connections. This is in contrast to a T-1 Internet port that could previously handle 20 consecutive VoIP calls and can now manage 50 or more because of compression. If you will need more than four analog ports in the next year, make sure that you have a second four-port card available. If you need more than eight analog lines and are running out of expansion slots, we recommend purchasing the 24-port card now and leaving the ports open until you need them.

Using analog interfaces

If you aren't versed in telephony jargon, we recommend using the Foreign Exchange Station (FXS) and Foreign Exchange Office (FXO) cards for service using regular phone lines, just as you do in your home. These lines come from your carrier and terminate in a small jack on the wall that are plugged into one phone. FXS interfaces connect directly to a handheld telephone or a dialer. The FXS cards provide the dial tone, caller ID, and ring voltage to the phone so that you understand that your call is being processed.

The FXO cards connect to the phone lines from your local carrier and transmit your call to the carrier for processing. These cards detect dial tone and ringing from the far end so that your FXS card can then forward this information to you.

The Digium analog cards allow you to either use four individual phone lines through standard RJ-11 jacks (the same jacks your home phone connects to) or an interface where you can use up to 24 phone lines through a specialized plug called an *amphenol connector* that separates every channel into 24 pairs of wires.

Analog cards from Digium are modular and can support FXO and FXS ports in any configuration. Your four-port card can have one, two, or three FXO cards with one, two, or three FXS ports. It's great to have options!

If you want 24 phone lines, you need special cabling with the amphenol connector and at least one more piece of hardware. You need a *break-out box* or *punch-down rack* to allow you to wire into the individual channels (in the case of the punch-down rack) or to plug your phones into one of the 24 standard RJ-11 jacks (in a break-out box).

You can purchase analog and digital cards that handle 1, 2, 4, and 24 lines as well as digital T1/E1/J1 cards for 1, 2, or 4 ports from the following companies:

- ✔ Voipsupply: www.voipsupply.com
- ✔ Sangoma: www.sangoma.com
- ✔ Digium: www.digium.com

Using external analog or digital cards with Asterisk allows you to connect to your existing phone lines. The larger T1/E1/J1 interface cards with two and four ports are only cost-effective in specials setups where you have an internal need for these phone lines or if you are reselling the lines to your customers.

We suggest that you expand your service with VoIP. Everyone in your company can access phone lines, without requiring you to purchase a single phone line for each employee. One dedicated Internet circuit can easily handle the voice requirements for 10 or 20 people with the correct configuration. Turn to the section "Sending calls out VoIP" later in this chapter if you decide to take your phone service to the next level.

Going digital and dedicated

Digium, Sangoma, and Zapata manufacture digital cards for Asterisk. Digium has the widest selection of digital cards that offer a two- and four-port model. Sangoma has a similar offering; its four-port cards are currently less expensive than the four-port Digium cards.

Sangoma offers a clear channel DS-3 card (capable of handling IP bandwidth equal to 28 T-1 lines) at this time. The company is also planning to release a channelized version of the card that can handle 672 voice channels (28 T-1s of 24 channels each).

The Zapata drivers are necessary and work well with any of the cards you choose. You have to love software that works with everything.

T-1 is an industry term for a circuit with 1.544 Mbps of bandwidth that is broken into 24 individual channels that can process one call each. This is the standard building block of dedicated digital telephony in the United States and Canada. Europe and much of the rest of the world use circuits called E-1s that are broken into 32 individual channels, giving you the capacity to handle eight more calls than the U.S. T-1 circuits.

All the cards are solid performers, but we have always experienced very good performance with the Digium cards. We also try to support our local Asterisk vendors, and because Digium resides in Huntsville, AL, where one of the authors runs his company, it seems the neighborly thing to do. If we take out the geographic bias, we recommend either the Digium or Sangoma T1/E1 cards.

Sending calls out VoIP

Asterisk comes standard with all the drivers you need to make VoIP calls. As long as you have a dedicated Internet connection and a NIC (network interface card) in your computer, you are ready to go. The only remaining piece of the puzzle you need is the ztdummy driver to help maintain the clocking on the calls, and that is only if you are transferring the calls internally within the Asterisk (or across to another Asterisk server) or if you are setting up conference calls. If you are dialing from a VoIP phone to a VoIP provider, you don't need the ztdummy driver.

Using VoIP to its fullest requires some research on your part. If you want to send and receive VoIP calls, you may need to have two VoIP carriers. Some carriers only provide *inbound* service (where people call you). This restriction allows the VoIP carrier to dodge the current federal requirement of providing 911 service. Other VoIP carriers specialize in *outbound* service (where you dial out) but may have limitations on some of the services provided. You may incur an additional charge for 411 and 911 services, as well as for a directory listing in the white pages. Even if you do get your name in the white pages, you may have to jump another hurdle trying to get your name in the 411 directories. Research these features with your carriers to be sure that you choose the correct carrier.

Communicating with your phones or dialers

So far, we have been speaking about hardware required to connect your Asterisk to a telecom carrier. The connection may be to a single analog telephone line or as many as four individual dedicated digital T-1 or E-1 lines. This is a vital link in sending calls to, or receiving calls from, the outside world, but it isn't everything you need. Unless you have a recorded message to a list of phone numbers in the same server that is running your Asterisk software, you need some type of interface to individual phones or another server that has a dialing program built on it (if you are a telemarketer).

Figure 1-1 depicts the possible setup of a server running Asterisk with an analog, digital, and dedicated Internet connection to individual carriers and wired into phones in the office on the other side. Its connections to telephones within your office are on the left, and the connections to a local carrier, long-distance carrier, or Internet Service Provider are on the right.

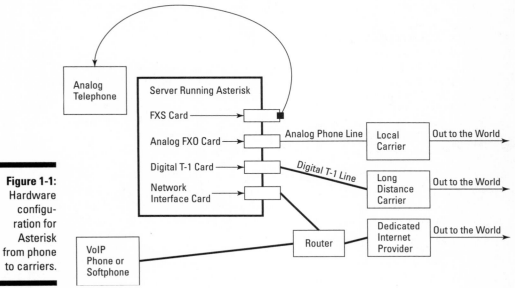

Figure 1-1:
Hardware configuration for Asterisk from phone to carriers.

The telephones you connect to your Asterisk server must be compatible with the type of telephony you are using. You can't use an Integrated Services Digital Network (ISDN) phone on a connection sending VoIP, just as you can't use a TDM phone on a digital circuit.

Several manufacturers produce VoIP phones that range in price from $50 to $600 per phone. Your specific application and budget dictate which phone is the best for you. Check out the following Web sites for VoIP phone manufacturers:

- VoIP Supply: www.voipsupply.com
- Cisco: www.cisco.com
- Linksys: www.linksys.com
- Polycom: www.polycom.com
- D-link: www.dlink.com
- Grandstream: www.grandstream.com
- Sipura: www.sipura.com

Chapter 2

Installing Asterisk

· ·

· ·

*T*he Asterisk installation process is straightforward, just like the installation of most software. We do recommend that you pay attention to the steps; by following the sequence, you reduce the chance for challenges down the line.

This chapter walks you through the simple procedure of identifying the hardware and software you need to install your Asterisk. The bits and pieces vary a bit, depending on what you plan to do with your Asterisk, but don't worry; we cover all the bases for you.

Taking Inventory of Your Hardware and Software

Asterisk is a real-time program that requires the full and undivided attention of your server. Whether you are using the software as a phone system for your business or as a basis for the value-added telecom service you are providing, you must have it on the best hardware possible. Regardless of the size of your company or server, the Asterisk program must be given top priority in all tasks to reduce the chance of network delays degrading the quality of your calls. A modest amount of delay caused by other network activity can result in static or sections of calls being dropped. At the other end of the spectrum, large delays could result in static or the failure to connect both inbound and outbound calls.

If you are using Asterisk for a small office of ten lines or less, we suggest the following minimum system requirements:

- ✔ 1GHz server.
- ✔ 1GB of RAM.
- ✔ Motherboard that supports Redundant Array of Independent Disks (RAID) on-board or an external RAID card for Serial Advanced Technology Attachment (SATA) drives.
- ✔ A pair of 400GB hard drives in a RAID setup to provide complete redundancy while providing capacity for the future. Hard drives aren't that expensive, so you should overengineer this part of your server to allow for growth.

If you're using voice mail, remember that one voice mail generally occupies between 300K and 1.2MB of memory. It is common to have 20 voice mailboxes; if you estimate 50 voice mails at 1.2MB, you're using 2GB of storage space. That being said, the 400GB hard drives should be more than sufficient for your needs.

We have witnessed over 40 consecutive VoIP calls processed at one time on a 60-MHz box with no problems (with the audio portion of the call bypassing the server and connecting directly between the origination and termination phones).

Larger applications that run more than 80 consecutive VoIP calls at a given time need a bit more power. In this case, you need a server consisting of the following features:

- ✔ Dual 2.8-GHz Xeon processors or better
- ✔ Minimum of 2GB of RAM
- ✔ A pair of 400GB hard drives with 160GB or better dual RAID 1 configuration with mirroring

Sharing the load with multiple servers

Asterisk is typically configured as a stand-alone server, but it really performs at impressive levels when it is joined to other Asterisk boxes in a cluster. The potential of Asterisk is truly realized when several Asterisk servers are clustered together. Just like any team, the output of the group is always more than the sum of its parts.

You can link Asterisk servers in several ways; some of the more common ways are as follows:

✔ **Dialplan switching inside the Asterisk system:** This option uses a `switch` statement in Asterisk to use a second server to look up the destination extension. This technique helps lighten the load of the server, but if one of the servers is down or offline, it can cause a delay in completing the calls. In the event that the server isn't active, you lose the time it took your system to realize that the server is down, creating a degree of congestion.

✔ **DUNDi (Distributed Universal Number Discovery):** This is a custom lookup application that polls Asterisk boxes in the network to find an active extension before defaulting to the main termination carrier. The process is efficient and may be perfect for your depending on your application.

✔ **Session Initiation Protocol (SIP) Express Router (SER) program:** This program allows you to take incoming VoIP calls and route them among multiple Asterisk servers. In this instance, the SER program either runs on a separate server or the same server as your Asterisk system. The SER is the first system to handle incoming calls and uses its routing process to handle all SIP requests and registrations. By acting as the first point of contact, the SER can direct the calls to an array of servers running Asterisk as well as have the option to terminate the calls to other hardware.

The biggest concern of implementing multiple servers in a VoIP application is ensuring that every server has full knowledge of registered end users and routes. This information is easy to establish with Asterisk on a single server, but it requires a separate database server or specialized protocol to share the information to all the Asterisk systems in a cluster.

Living with Linux

Linux is the required operating software for Asterisk. You must have a good working knowledge of Linux to install and work with Asterisk because all the installation and programming is done within that framework. You need root access to install Asterisk and run certain programs, such as aspects of the voice-mail script.

You should not set up Asterisk to run on the root directory of Linux (which could cause security concerns). Set up a useraccount and usergroup that are limited to only the permissions required to run the software.

Appendix C covers the basics of Linux to get you up to speed.

Downloading the Asterisk Software

Before you can install anything, you must have it available. In Chapter 1, we give you a rundown of the software drivers necessary to make a VoIP or time-division multiplexing (TDM) call. In this section, we give you the specific software and the sites where you can download it.

Install all hardware prior to installing any software. The server identifies the Digium, Zapata, or Sangoma cards during the installation of the Asterisk software. Installing the cards prior to compiling and establishing the software is just easier than trying to integrate them after the fact.

The Asterisk software is constantly evolving. Because of this, many revisions are available at any given time. With all the choices, you must decide which one is right for you. Luckily, the many revisions of software are divided into the following two categories:

- ✔ **Stable:** Stable versions have been released and out in production long enough that all the bugs have been found and worked out. If you are a first-time user and aren't intimately familiar with the software, stick with the stable version.

- ✔ **Head:** Head versions are brand-new releases and a viable alternative if you're more familiar with the ins and outs of the Asterisk software and know its temperament. New releases come out all the time, and using a head version that is seven to ten days old gives you the latest and greatest features with only modest exposure to bugs.

The head version is not a solid release. You should walk out on the thin ice of unchartered software only if you feel confident enough to troubleshoot it if something goes south.

Retrieve the source code for Asterisk from the Digium FTP site located at `ftp://ftp.digium.com/pub/telephony/asterisk` and save it on your Linux server in the /usr/src folder. The source code is listed numerically with each new release. The file you're looking for begins with `Asterisk-1`, has the version number in the center, and ends with `.tar.gz`. This site has other useful files and drivers that you need to fully install Asterisk, so keep the site bookmarked.

Source code is a programming file that must be assembled (or *compiled*) into an executable program. Less complex software programs like WinZip or Textpad simply require downloading and a one-click installation. They are already fully built, compiled, and ready to go. Software that uses a source code differs in the fact that you must retrieve the software, along with all related supporting software called *libraries,* and compile it into an executable format. Only after you have compiled your software can you then install it on your system.

In addition to the source code, you need to download these files:

✔ `asterisk-1*.tar.gz`: This file contains the basic Asterisk software.

✔ `asterisk-sounds-*.tar.gz`: This file contains the standard Asterisk sounds.

This is where all the useful telecom sounds and recorded messages are kept, such as "Please enter the extension number of the person you wish to speak to" and "For a dial-by-name directory, please press the first few letters of the person's last name." The sounds directory also includes additional custom phrases and words as well, so check it out when you have some free time.

✔ `zaptel.-*.tar.gz`: The Zaptel drivers you need if you use your Asterisk over standard analog phone lines.

✔ `libpri-*.tar.gz`: This is the Primary-Rate Interface (PRI) library software that allows you to use Integrated Services Digital Network (ISDN) lines from your carrier.

You must use the ztdummy driver for timing if you do not have a Zaptel card (POTS/Digital T1) but want to do conferencing, InterAsterisk eXchange (IAX), or anything else that may require a timing source. In Linux 2.6, the ztdummy module uses the internal kernel timing source; in version 2.4 kernels, you must load a USB module as a pseudotiming source. You can see the details at `www.voip-info.org/wiki-Asterisk+timer+ztdummy`.

If you are using MySQL for database access from Asterisk, we also recommend that you download the asterisk-add-ons software at `http://ftp.digium.com/pub/asterisk/releases/asterisk-addons-1.2.3.tar.gz`.

Getting a head version of Asterisk

You can access the head version of Asterisk in a couple of ways. The most basic is by downloading the source code from the Digium site at `www.asterisk.org/download`. The most current release is listed in a column on the right side of its Web site listing the files and a release number. The head versions of the software are easy to identify because they have `beta` in their names. Currently, you can download the 1.4.0-beta release for the Asterisk source code, Asterisk add-ons, libpri, and Zaptel kernel interface.

You can also retrieve the source code with a program called Subversion. Most Asterisk customers use this program to download the entire source code before compiling and installing. Many versions of Linux include some sort of packaged installation program. For those Linux distributions that don't have an installation program, you can download Subversion from http://subversion.tigris.org/.

When you have Subversion installed on your computer, use the following code on the Linux server that you want to receive the Asterisk software:

```
cd /usr/src
svn checkout http://svn.digium.com/svn/asterisk/trunk
        asterisk
svn checkout http://svn.digium.com/svn/zaptel/trunk zaptel
svn checkout http://svn.digium.com/svn/libpri/trunk libpri
```

Securing a stable version of Asterisk

We show you how to download a head version of the Asterisk source code with Subversion earlier in the chapter; see the previous section to find out how to use Subversion.

You can also use the wget software to grab the stable file.

wget is common file-transfer software that comes standard with most Linux packages. The software allows you to easily download files via HTTP or FTP via the console. Because you need .gz compressed files from the FTP site, wget is good choice.

Download the source code from the /usr/src directory on your server. Using the asterisk (*) in the command ensures that you're downloading the most recent release of the software. If you are using wget, type the following code at the root directory:

```
# cd /usr/src
#wget - -passive-ftp
        ftp.digium.com/pub/asterisk/asterisk-1*.tar.gz
# wget - -passive-ftp
        ftp.digium.com/pub/asterisk/asterisk-sounds-
        *.tar.gz
# wget - -passive-ftp ftp.digium.com/pub/zaptel/zaptel-
        *.tar.gz
# wget - -passive-ftp ftp.digium.com/pub/libpri/libpri-
        *.tar.gz
```

Building a tarball

After you download the source code, you need to extract it. We recommend using the GNU `tar` application, a tool that takes multiple files and pulls them together into a single *tarball* (archive). If you used the `wget` software to transfer the files to the usr/scr/ folder, you're extracting the files to the same place.

Place all files in the /usr/src directory. This is the normal location to store user source files that remain on the system. Execute the `tar` application by entering the following code from a root access command line in Linux:

```
# cd /usr/src
#tar zxvf asterisk-*.tar.gz
#tar zxvf asterisk-sounds*.tar.gz
#tar zxvf zaptel-*.tar.gz
#tar zxvf libpri-*.tar.gz
```

The executable files are placed in their corresponding directories after they are compiled.

The `zxvf` located in the second position of each line of code is hacker shorthand. The syntax breaks down as follows:

- `z`: Identifies that the file is to be unzipped/zipped (when creating an archive)
- `x`: Identifies that the files within it are to be extracted
- `v`: Tells Linux to view the files
- `f`: Specifies the next parameter in the filename

Compiling Asterisk

After you compile the Zaptel drivers, the libpri, and the ztdummy (if you need it), you can bring together all of the constituent elements and compile Asterisk itself.

We recommend the following sequence for installing the Asterisk source code and the supporting libraries and drivers:

- libpri source file
- Zaptel drivers
- Asterisk source code
- asterisk-sounds file
- Asterisk add-ons

The process for compiling each software component is the same. They all begin with a command to select the file in the specific folder you wish to compile.

Starting with libpri

To compile the libpri file, type the following code at the command line:

```
# cd /usr/src/libpri.1.2.3
# make
# make install
# cd
```

Change the `libpri.1.2.3` to the exact version of the libpri software you downloaded.

The `make` step reads the `Makefile` in the current (or specified) directory to gather the parameters of the file and prioritize them. It then compiles each section individually until this step is complete. The `make install` command takes the compiled code and installs it in the default directory location for the application and resources. The last line, the `# cd` command, sends you back one directory so that you can type your next command.

Moving to Zaptel

To compile the Zaptel drivers, type this code at the command line:

```
# cd /usr/src/zaptel.1.2.6
# make
# make install
# cd
```

Be sure to change `zaptel.1.2.6` to the specific version of the Zaptel driver you downloaded.

Bringing up Asterisk

Install Asterisk by typing the following code at the command line:

```
# cd /usr/src/asterisk-1.2.9.1# make clean
# make mpg123
# make
# make install
# make samples
# cd
```

The `make samples` command copies the sample configuration file in the Asterisk source files and places it in the `/etc/asterisk/directory` folder. The sample configuration files allow you to see generic optioning for a variety of configurations. If you want to see a basic dialplan for implementing conference rooms or voice mail, these sample files give you a template to work from.

The `make mpg123` command establishes the infrastructure necessary for Asterisk to provide music on hold. The distribution of Linux you're using determines the necessity of this file. It works with Slackware, some Debian, and current releases of Gentoo, but it may not be compatible with older Linux distributions. If you don't want or need it, simply bypass this step.

Turning on sounds and add-ons

You don't need to compile the Asterisk sounds and add-ons, but we recommend doing so, simply to make the installation complete. Type the following code at the command line:

```
# cd /usr/src/sounds.1.2.1
# make install
# cd
```

```
# cd /usr/src/addons.1.2.3
# make install
# cd
```

Overcoming Common Compiling Issues

One beauty of Asterisk is that it installs easily. You rarely have problems compiling the software unless you're missing common libraries in your Linux distribution. The system either provides you with a list of errors after you compile the software or simply presents another command prompt. If you have the command prompt, you're free to configure your hardware.

Two common errors are defective or missing files:

- **To remove a compiled file:** Execute this command in the directory with the defective files:

  ```
  # make clean
  ```

- **To install and recompile a missing library file:** Execute this command:

  ```
  # make
  ```

If you can't find the problem, you may have to wipe out all your compiled files by running the make clean command and begin the installation from scratch. If this doesn't resolve the problem and it isn't compiling without errors, we suggest getting professional help. In this case, that means speaking to a competent Linux technician or, in rare cases, a Linux engineer with software development experience.

Managing Server Security

Asterisk is a server program just like any other, and it can fall prey to intentional and unintentional attacks. The thought of an unintentional attack does seem strange. How can you be minding your own business and suddenly realize that you have just invaded a server? It happens all the time.

Unintentional attacks are called *rogue Real-Time Transport Protocol (RTP)*. The name refers to the voice portion of a VoIP call that is accidentally directed to a random server or is looped to terminate at the same server repeatedly. This problem is frequently the result of incomplete or incompatible VoIP software attempting to reach you. The constant barrage of calls hitting your server probably won't crash it, but they can slow the server considerably.

Intentional attacks are called *denial-of-service (DoS)* attacks. DoS attacks happen when your server is being overwhelmed with call attempts from a remote device, preventing it from providing service to your legitimate customers.

You can also have your server compromised by flaws in the code (new or old) that others may exploit. You have no way to proactively prevent these problems except for keeping up with the latest version. Code flaws are just that: flaws in the programming. Although we are sure that you are taking every precaution to prevent unwanted attacks and control of your server by hackers from the outside, denial-of-service attacks can take place.

Opening the management port (telnet) publicly can open a security hole, but these holes are only as vulnerable as the software that opens them. Don't open the port unless you are comfortable accessing and securing it. The standard Asterisk configuration disables the port by default, so you need to enable it before you design Asterisk. You can use the telnet port by opening it for local connections only. Then you can set the port as a secure shell (SSH) connection using the SSH client. This client is linked from a specified port to the telnet port on the Asterisk. The specific configuration for the connection depends on the specific SSH client program you are using, so unless you are comfortable working with SSH, you might want to skip this option.

A lack of security on your server allows malicious hackers to gain control of your hardware and run rampant. Unauthorized control happens when you have insecure passwords, allowing your server to be open to the world, or when you take no security measures to limit connections to server applications on your computer. All of these actions or inactions allow the user to control the server in any fashion (whether it be making a call or shutting the server down).

The only way to prevent problems is to secure your server as much as possible. This can be as easy as putting your server behind a firewall or running an application called iptables on your server to limit the Internet Protocol (IP) addresses from which your server can accept connections.

This setup and configuration are more involved that just setting up Asterisk. You must consider all aspects of your configuration when taking these measures, because it is easy to protect your server so much that you reject all traffic being sent to it. Having a secure system is nice, but your server must also be permeable enough to be used.

If you are planning to run a publicly available Linux server — whether you are running Asterisk or some other server program — you need to know your security options. Installing a firewall and being aware of all available software to prevent undesirable connections are necessary requirements for you to successfully manage a public Asterisk server. A server available to the Internet means that it is available to every hacker, client, and carrier, both good guys and bad guys.

Chapter 3

Installing AsteriskNOW

*T*his chapter shows you how to download the AsteriskNOW software and then how to install it on your Linux server. If your server isn't using Linux as the operating system (OS) when you start, it will be by the time you finish the installation.

Chapter 8 shows you how to configure AsteriskNOW after you have it installed on your Linux server.

Downloading AsteriskNOW

The AsteriskNOW software is available for free; you can download it from www.asterisknow.org. The Web site prominently displays the AsteriskNOW free download link that you click to download the software.

Before downloading the software, you must log in to the Asterisk site. If you've never been to the Asterisk Web site, you need to create an account (click the Create an Account link).

After you log in to the site, you can download a version for either 32- or 64-bit hardware. Simply choose the link for the classification that matches the server on which you're going to load AsteriskNOW and begin the download process. The file you download is called asterisk-1[1].4.0-x86-disc1.iso.

The file download is presented in a pop-up window, so you may need to adjust your pop-up blocker to allow the file to download. If the pop-up window doesn't even appear, either disable your pop-up blocker temporarily, or select the AsteriskNOW Web site as an approved site for pop-ups.

Don't download the software directly to your Linux server. Instead, the software is designed to be burned to a CD and used as a boot-up disc for your Linux server.

After you download the software, place a blank CD in your disc drive and double-click the `asterisk-1[1].4.0-x86-disc1` file. The file searches for the CD-burning software on your computer and begins writing the data to disc.

Allow your CD-burning software to finish its job, and in a few minutes, you have a CD capable of loading AsteriskNOW and Linux on a server. Remove the CD from the burner and label it AsteriskNOW so that you can identify it later.

Before you load the software onto your server, first install all the analog, digital, and IP hardware in the server. The CD polls your server to find these devices, allowing a much easier installation after the hardware is established. See Chapter 2 to find out how to install all the hardware you need.

Booting the AsteriskNOW CD

The CD you've created is used to boot the computer and isn't loaded on a live machine. Power up your server and immediately place the AsteriskNOW CD into the CD drive.

As the server is booting, press F10 to prevent it from booting in a normal manner from the hard drive. You may have to select your CD drive to have your server boot from it, or the server may automatically search for the disc. The servers we've loaded with the AsteriskNOW software immediately jumped to the disc drive and displayed the AsteriskNOW installation window.

Don't load the CD on a PC you want to use for your desktop, such as your MS Windows machine. It will wipe out all data and change the Master Boot Record.

You see the following command at the bottom of the first installation screen:

```
boot:__
```

Your cursor is blinking with anticipation to the right of the command. Press Enter, and you're off and running.

The Welcome to AsteriskNOW screen appears with some general information about AsteriskNOW; your options to use a mouse, keyboard, or keyboard shortcuts to navigate the installation; and a Help window on the left of the screen.

You can collapse the Help window if you find it distracting. We find it, well, helpful if you have questions about the options available on the installation screens.

You take these steps to complete an installation:

1. Choose an installation.
2. Identify how to partition your drive.
3. Identify any network devices.
4. Locate your time zone.
5. Assign a password.
6. Reboot your server.

We cover each of these steps in the following sections.

Choosing your variety of installation

On the Installation Type screen, you choose what kind of installation you want. You have the following three options:

- **Express Installation:** This is our preferred method of installing the software. It establishes all the basics you'll need, and it's the method we use in this chapter.

- **Custom Installation:** This option allows you to modify how the basic software is installed, but Asterisk still handles most of the rudimentary elements of installation for you.

- **Expert Installation:** If you choose this installation, you could probably run the whole installation from the command line in Linux.

We recommend choosing the Express Installation option. Click the Next button to continue to the next screen.

Identifying your partition preference

On the Automatic Partitioning screen, you choose how to partition your drive. Your computer's hard drive has partitions that prevent data stored in one section from overtaking other sections reserved for use by another system. This is important because if you remove the partitions, you over-write all of the data and operating software for the entire computer. To make a long story short, all of your existing files on the computer are gone. You have the following partitioning options:

- ✔ **Remove All Linux Partitions on This System:** This option tells the installation to acknowledge the partitions and distribution (distro) of Linux on your server, only as far as the real estate it occupies on the machine. After identifying the disc space used by Linux, the AsteriskNOW software then replaces it with the version of Linux on the CD.

- ✔ **Remove All Partitions on This System:** This option allows the installation of the new Linux distro to replace any and all operating systems partitioned on the server. If you're wondering what this really means, imagine reformatting your hard drive. That's essentially the power you give the installation if you choose this option.

- ✔ **Keep All Partitions and Use Existing Free Space:** This is the safest option, but it could result in a failed installation if you run out of space.

Because AsteriskNOW is a real-time application, we don't recommend running it on a machine with any other active OS. We suggest you choose either the first or second option and allow the installation to replace the existing OS.

After you select the partition option, click the Next button.

Identifying network devices

The Network Configuration screen auto-locates your network interface card (NIC). You have options to set your hostname, either automatically via Dynamic Host Configuration Protocol (DHCP) or manually by inputting the host domain URL, but the Express Setup already selects this for you.

The Network Devices area identifies the NIC found on your server. You can edit the NIC if you want to assign a different IP address to it, but the DHCP is used by most servers to automate this process so the default IP identified should be all you need.

If AsteriskNOW doesn't find your NIC, consult your Network Administrator or a Linux technician. The NIC card may not be compatible with Linux, or some bit of coding may be required.

The DHCP also provides the hostname used for your network device, which is why the Set the Hostname option defaults to Automatically via DHCP.

The miscellaneous settings for your gateway (the IP address from which you leave the safety of your network and hit the open Internet) and the Domain Name Service (or DNS that translates the hostname of mydomain.com into an IP address so people can connect to your server) aren't used in the Express Setup. Most IP providers utilize a primary DNS server and a secondary one in the case the first IP address is unavailable. AsteriskNOW allows you even more protection, allowing a tertiary DNS IP to be used.

Review the settings AsteriskNOW has set up and click Next.

Locating your time zone

The Time Zone Selection screen is where you choose what time zone you're in. Choosing your time zone is trickier that it looks. The drop-down box has every time zone in the world, but it isn't based on simple things like country and city. The designations are more akin to continent and city. This is why you'll find America/Lima as an option for a time zone directly above America/Los Angeles. Yes, they're both in the Pacific time zone, but they are also on opposite sides of the equator.

You can always opt for the point and shoot method of setting your time. Set your cursor on the yellow dot closest to your location and click it with your mouse. A pop-up window identifies the city you've selected.

Select the System Clock Uses UTC option if you want to use UTC as the basis for your timing. UTC is Coordinated Universal Time, a superior replacement to GMT (Greenwich Mean Time), which uses a complex science based upon the rotation of the Earth to maintain an accurate time within 0.9 seconds. The added benefit of UTC is that it accounts for Daylight Saving Time, regardless of where in the world you're located. The default is to use UTC, and we recommend keeping that box checked.

Select the correct time zone based on the most recognizable city listed in your time zone, and then click the Next button.

Assigning a password

After you assign the time zone for your server, you go to the Administrator Password screen, which requires you to assign an administration password to the server for Asterisk. You must enter the password twice to confirm that you spelled it correctly.

The password you assign at this stage of the installation is very important. Use a different password than the one you assigned for the root access of Linux. This password is the only way you can log in to the system after you install it.

Enter the password and click the Next button. Click the Next button in the About to Install screen, and the system goes to work. In 15 to 25 minutes — depending on the speed of your server — your installation completes, and your CD is ejected (on most servers).

Completing the installation by rebooting

Don't be fooled when the software says the installation is complete and the CD is ejected. You're not ready to program your Asterisk until after you reboot the server. Actually, the only option you're given when the installation is complete *is* to reboot your server. Many important things happen during the reboot: The files are uncompressed, and all cards and interfaces you've installed are identified.

You must have your server connected to the Internet while it's rebooting. The system isn't gathering information from the Internet during this time, but the software is identifying the public IP address that's assigned to your server. This information is essential in making the connection from the Web portal you'll use to configure your Asterisk.

Watch your monitor during the rebooting process. Every element of the hardware and software scrolls in individual lines of text during the reboot process. Any essential information that's unavailable is identified as [Failed] in red on the right side as the required module is scrolled up. As long as your server is connected to the Internet during the boot process, nothing should be listed as [Failed].

Arriving at the AsteriskNOW Console Menu

The AsteriskNOW Console Menu is the final destination of your server after rebooting. It provides you with a list of options as well as a vital piece of information: the IP address that provides the graphical user interface (GUI) for programming your Asterisk. The verbiage of the window is as follows:

```
To configure the system, point your browser to:

http://192.168.1.561
```

Of course, your specific IP address is different than ours, so write down your IP address and keep it in a safe place. You need it when programming your Asterisk in Chapter 8.

The second window presented in the AsteriskNOW Console Menu enables you to do the following:

- Update the system
- Enter or return from the command-line interface (CLI) for Asterisk
- Restart the computer
- Shut down the system
- Reboot the system
- Exit the AsteriskNOW Console Menu

The most helpful option is the one that allows you to access the Asterisk CLI. Use the following key combinations:

- To enter the CLI, press Alt+F9.
- To return to the AsteriskNOW Console Menu, press Alt+F1.

Chapter 4

Configuring the Hardware

* *

* *

You've gone through all the hard work — you've installed the hardware and downloaded, extracted, and compiled the software — and now you want to make a phone call. Not so fast! You still have to set up all the cards and interfaces with the correct telephony configuration so that they can effectively communicate with everything around them.

In this chapter, we cover how to configure the individual analog and IP ports, as well as tell you how to build connections to other Asterisk servers. These connections will establish the paths over which your calls will be sent, so it is important to have these set up correctly. Every connection has many options available, and we discuss these options to ensure that you get the most from your Asterisk.

Configuring Everything You Need

You should configure your Asterisk server as you need it. A specific sequence isn't required to make the whole thing work. Asterisk does have a variety of configuration files available to you depending on your application. These files are located in the following directory:

```
/etc/asterisk
```

We show you how to configure every aspect of Asterisk; skip anything that doesn't pertain to you. For example, loading up on Session Initiation Protocol (SIP) ports doesn't make sense if you aren't using VoIP. You don't have to configure analog circuits if you aren't going to use them. Similarly, you don't need to configure IP connections if you aren't going to use VoIP. We do suggest that you configure at least one connection of each type so that they're ready in case you need them in the future.

You must reload any configuration file you have modified before the changes take effect. You don't want to go through the effort of building a connection and then be frustrated because the feature still isn't working. So remember to reload!

Setting Up Your Zaptel Cards

The configuration for Zaptel cards comes with several options, most of them specific to the carrier or phone system that connects to your Asterisk. These options include the ability to modify the transmit (`tx`) and receive (`rx`) levels on your calls to isolate and eliminate line noise (such as static). Most of the added bells and whistles aren't that necessary. The items you do need are as follows:

- **Context:** This specifies where incoming calls are sent.
- **Switchtype:** This is only for ISDN circuits and identifies the type of local carrier central office switch into which you are interfacing. Your options are

   - **National:** National ISDN 2 (default)

   - **DMS100:** Nortel DMS100

   - **4ESS:** AT&T 4ESS

   - **5ESS:** Lucent 5ESS

   - **euroisdn:** EuroISDN

   - **NI1:** National ISDN 1

   - **qsig:** Q.SIG

- **Signaling:** This is for connections to analog or digital phone lines (non-VoIP) and identifies the signaling protocol being used. It could be `loopstart` or `groundstart`, for example.

- **Group:** This specifies a collection of analog or digital channels used to receive calls. The group causes a single channel to be selected from the group whenever you dial in on one of the TDM (non-VoIP) lines.

✔ **Channel:** This specifies the individual channels on the TDM device that the options apply to. Each phone line represents a single channel, so a card with two analog lines would have two possible channels. Similarly, a digital T-1 card would have 24 available channels.

Other options are available to you in the Asterisk configuration. The need to modify these settings varies based on how your phone lines are delivered to your Asterisk as well as the features available from your carrier:

✔ **HDLC:** This is only used when you have a digital ISDN T-1 circuit partitioned for regular voice calls on one section and data transmission over the rest of the circuit.

✔ **Showing caller ID:** Do you want other people to see your caller ID?

✔ **Enabling three-way calling:** This allows you to conference another person on an existing call or use some Asterisk transfer features.

✔ **Tx gain/rx gain:** These features allow you to change the electrical strength on the line, thereby adjusting the voltage in the event that the signal strength coming into your Asterisk box is low. Copper wire has inherent resistance in it, which reduces the strength of a signal passing over it. The longer the cable, the more signal is lost. The world of telephony is aware of this and boosts, or *regenerates*, the signal at specific points.

Building Analog Connections to Your Carrier

If you are using the Asterisk private branch exchange (PBX) for your office or as the basis of your business, you probably need at least one analog phone line (just in case your digital or Internet lines are down). The phone line that connects your Asterisk to your local carrier is referred to as a *Foreign Exchange Office (FXO) line,* and you need one FXO interface port for every analog phone line you want to have access to from your Asterisk server.

FXO modules are configured with Foreign Exchange Station (FXS) signaling. This idea may seem counterintuitive, but look at it closer. The modules are named based on the hardware they connect into. That is why a connection to the local carrier office requires an FXO module. In the case of FXO modules, the server is acting as an end station, receiving dial tone and ringing; hence it's configured with FXS signaling. Similarly, an FXS module is connected to an analog phone (station) but functions to generate ringing and dial tone — so it acts like a local carrier office — and it's configured with FXO signaling.

Just because you receive analog lines from your carrier doesn't mean that you have to use analog phones. Similarly, just because you have analog phones at your office doesn't mean that you are limited to using analog phone lines from your carrier. Asterisk offers wonderful and dynamic software that easily converts calls from analog to VoIP, or VoIP to Integrated Services Digital Network (ISDN). You have the flexibility with Asterisk, so use it!

Don't plug an analog phone line from your local carrier into an FXS module; analog phone lines must go into the FXO cards. The unexpected voltage being sent down from your carrier to deliver ringing and dial tone could destroy your FXS module.

Configuring the FXO card with FXS signaling

To configure the FXO card, you need to modify the `zaptel.conf` file located in the `/etc/` directory. You can confirm that the file is available by using the following command:

```
ls /etc/zaptel.conf
```

After you find the file, use your standard Linux editor to modify it.

After identifying the numerical position of the FXO port on your analog card (position 1, 2, 3, and so on), you're ready to configure it. This is how you would configure an analog card with the FXO port in position 1 and functioning to receive dial tone and ringing from the local carrier for FXS signaling:

```
fxsks=1
loadzone=us
defaultzone=us
```

Establishing port type and signaling

```
fxsks=1
```

The first line of the code identifies the type of interface on the circuit: `fxs` for Foreign Exchange Station, `ks` for the signaling to indicate Kewlstart, and `=1` to identify the analog port you're configuring.

Analog lines within Asterisk have the following three signaling options:

✔ **gs:** This option is for groundstart circuits. These circuits aren't typical, but some still exist. If you have new phone service, you probably don't have groundstart lines.

- ✔ ls: This option is for loopstart circuits. You find these circuits in residential phone lines.

- ✔ ks: This option is for Kewlstart. These circuits function with standard loopstart lines, but allow you more functionality.

Determining your zones

```
loadzone=us
defaultzone=us
```

The loadzone identifies the telecom sounds particular to your nation. Setting it to loadzone=us enables all the ringing, busy signals, and standard telephony sounds of the United States to be attributed to the FXO port.

If you don't identify a loadzone for the channel, don't worry. As long as you have identified the defaultzone somewhere in the system, Asterisk uses it to generate the required sounds and noises.

Some of the available loadzones are as follows:

- ✔ #loadzone=us: United States

- ✔ #loadzone=es: Spain

- ✔ #loadzone=it: Italy

- ✔ #loadzone=fr: France

- ✔ #loadzone=de: Denmark

- ✔ #loadzone=uk: United Kingdom

- ✔ #loadzone=fi: Finland

- ✔ #loadzone=jp: Japan

- ✔ #loadzone=sp: Spain

- ✔ #loadzone=no: Norway

The loadzones are based upon the standard International Organization for Standardization (ISO) country codes. You can find the entire list on its Web site at www.iso.org/iso/en/prods-services/iso3166ma/ 02iso-3166-code-lists/list-en1.html. If you don't want to type all that, simply search for *ISO country codes* on your favorite Internet search engine and you'll probably end up in the same place.

Also set your defaultzone. It is always going to be the same as the loadzone of your program, unless you have a special application where you need a group of extensions to ring as if they are in a different country. If you're in the United States, set your defaultzone as

```
defaultzone=us
```

Identifying your driver

After configuring the individual ports of your interface cards within Asterisk, load the driver for the card. Every card needs a specific driver to make it work. Table 4-1 provides a quick reference for drivers you need for specific hardware. You only need one driver per card, so match up your hardware to the appropriate driver.

Table 4-1	Software Drivers for Accessory Cards	
Interface Card	*Driver*	*Command*
Pciradio cards using Zaptel driver	pciradio	`modprobe pciradio`
Older ISA T1 cards	torisa	`modprobe torisa`
Older T1 cards	tor	`modprobe tor`
Certain port cards	wcfxo	`modprobe wcfxo`
Single-port T1 cards	wct1xxp	`modprobe wct1xxp`
Two- and four-port T1 cards	wct4xxp	`modprobe wct4xxp`
Certain port cards	wctdm	`modprobe wctdm`
24-port analog cards	wctdm24xxp	`modprobe wctdm24xxp`
Newer T1 cards	wcte11xp	`modprobe wcte11xp`
USB driver for use by ztdummy	wcusb	`modprobe wcusb`
Generic driver loaded after all others	zaptel	`modprobe zaptel`
HDLC Ethernet data over analog	ztd-eth	`modprobe ztd-eth`

If you're not sure what card you have, the Linux command `lspci` identifies your Digium or Sangoma card type, as well as the manufacturer. Execute the command as follows:

```
# lspci
```

The command lists the card type and manufacturer, similar to the following information provided for the Zapata card by the Tiger Jet Network:

```
00:1f.3 SMBus: Intel Corporation 82801DB/DBL/DBM
        (ICH4/ICH4-L/ICH4-M) SMBus Controller (rev 01)
00:1f.5 Multimedia audio controller: Intel Corporation
        82801DB/DBL/DBM (ICH4/ICH4-L/ICH4-M) AC'97
        Audio Controller (rev 01)
01:00.0 Network controller: Tiger Jet Network Inc.
        Tiger3XX Modem/ISDN interface
01:08.0 Ethernet controller: Intel Corporation 82801DB
        PRO/100 VE (LOM) Ethernet Controller (rev 81)
```

The lspci command identifies your hardware by the actual chipset used to power the cards, not by the manufacturer that's printed on the box. The Tiger Jet Network Inc. card, for example, is specific to Zapata cards because that card uses that chipset. Some cards that are not Zapata cards may use this same chipset. You need to be familiar with the hardware in your server and know what hardware drivers you need.

Installing the driver

In spite of the number of cards and commands listed in Table 4-1, installing the right driver isn't that complicated. You need to know the following information to install the right driver:

- ✔ The number of ports your card has
- ✔ Whether your card is analog and digital
- ✔ The generic Zaptel driver

For example, if you have a 24-port analog card, simply execute the following commands from the /etc/ directory:

```
modprobe zaptel
modprobe wctdm24xxp
modprobe wcusb
```

If you have a type of card that isn't listed in Table 4-1, you can identify the driver you need by looking through the ReadMe file in the Zaptel source directory. Scan the document until you find your card, and the name to the far left identifies the driver you need.

Here's how our ReadMe file looks:

```
tor2    T400P - Quad Span T1 Card
        E400P - Quad Span E1 Card

wct4xxp   TE405P - Quad Span T1/E1 Card (5v version)
          TE410P - Quad Span T1/E1 Card (3.3v version)
```

```
wct1xxp    T100P - Single Span T1 Card
           E100P - Single Span E1 Card

wcte11xp   TE110P - Single Span T1/E1 Card

wcfxo    X100P - Single port FXO interface
         X101P - Single port FXO interface

wctdm or    TDM400P - Modular FXS/FXO interface (1-4 ports)
wcfxs

wcusb or    S100U - Single port FXS USB Interface
wcfxsusb

torisa    Old Tormenta1 ISA Card

ztdummy    UHCI USB Zaptel Timing Only Interface
```

After you know the driver you need, simply replace the drivername with the required driver and execute the following command:

```
modprobe drivername
```

Validating your successful configuration

Before you run off assuming that everything is working correctly on the FXO port you just configured, checking it is a good idea. This is done with the ztcfg program that comes with the Zaptel drivers. Type the following code:

```
# /sbin/ztcfg -vv
```

This command displays all the ports and their configurations. For a Zaptel configuration, the display would look like this:

```
Zaptel Configuration

Channel map:
Channel 01:  FXS Kewlstart   (Default) (Slaves: 01)
1 channels configured.
```

Resolving a failed configuration

If the ztcfg program gives you anything other than a clean response, reload the correct driver from Table 4-1 and review your work with the ztcfg program.

Removing defective drivers

If you find that something has gone haywire in the installation of a device driver, remove the driver and try again. Removing a driver is done using the rmmod (remove module) program by typing either of the following command:

```
# rmmod wctdm
```

or

```
# modprobe -r wctdm
```

In this example, the wctdm portion of the command refers to the specific driver installed for the FXO card. After all the code is removed, start again.

Building an analog connection to your phone

The telecom world sees the port on your server that connects to the phone on your desk as an FXS port. This, of course, means that you have to configure the port with FXO signaling. After you have the FXO signal configured, you add one line of code to the zaptel.conf file to build the connections to your phone, as follows:

```
fxsks=1
fxoks=2
loadzone=us
defaultzone=us
```

The channel identifiers of 1 and 2 are specifically designated toward the channel number. You can efficiently configure a pair of FXS and FXO channels by changing the parameters to the following:

```
fxsks=1-2
fxoks=3-4
```

When you check the configuration of the zaptel.conf file using the /sbin/ztcfg -vv command, you get the following response:

```
Zaptel Configuration
========================

Channel map:
Channel 01: FXS Kewlstart (Default)    (Slaves: 01)
Channel 02: FXO Kewlstart (Default)    (Slaves: 02)

2 channels configured.
```

Tying the analog cards to Asterisk

Until you link the Asterisk to the configured hardware, the server thinks you're running some other application with the `zaptel.conf` file. You can rectify this situation by modifying the similarly named `zapata.conf` file that identifies the telephony parameters of hardware used by Asterisk.

The `zapata.conf` file is more involved than the `zaptel.conf` file you need to configure for the FXO card. This is because the FXO card interfaces with the hardware at your local carrier, so it must follow normal telephony protocols to send and receive information.

The `zapata.conf` file is an Asterisk file that embodies the greatness of the software: its flexibility. If you wanted a phone system that didn't give you options, you would have spent a lot more on a proprietary system. Before you tie the analog card to your Asterisk system, you must have two channels configured. If you're not sure how to configure your channels, see the previous section. Your file should look like this:

```
[trunkgroups]
; define any trunk groups

[channels]
; hardware channels
; default
usecallerid=yes
hidecallerid=no
callwaiting=no
threewaycalling=yes
transfer=yes
echocancel=yes
echotraining=yes
immediate=no

; define channels
context=incoming
signaling=fxs_ks
channel => 1

context=internal
signaling=fxo_ks
channel => 2
```

The `[trunkgroups]` section is for NFAS/GR-303 connections, which is an estranged variety of ISDN. We have never seen this used in a dialplan, so it's safe to ignore it for now.

Selecting features in the channels section

```
[channels]
; hardware channels
; default
usecallerid=yes
hidecallerid=no
callwaiting=no
threewaycalling=yes
transfer=yes
echocancel=yes
echotraining=yes
immediate=no
```

The `[channels]` section of the `zapata.conf` file lists all the features available to you on your configured channels. When you have the FXS and FXO ports configured, your standard options are as follows:

- `usecallerid=yes`: This is a pretty standard feature that allows you to read the caller ID sent on incoming calls.

- `hidecallerid=no`: This is the flip side of using caller ID. By selecting `hidecallerid=yes`, you effectively block your caller ID so that your calls appear as private.

Some people don't like to receive calls from people who don't display their caller ID. To make some money off this niche market, the local carriers generally offer a service that rejects calls placed to your phone number without a valid caller ID. We suggest that you always display your caller ID, unless you have a good reason for blocking it.

Your local phone carrier may offer features that allow you to block or unblock your caller ID called *star codes*. The most common star-code feature is *69, which calls the last person who called you. You must build the star codes into your dialplans just like you do for any extension. In the absence of a *69 extension, Asterisk simply sends a call where *69 is dialed to its internal extension *69.

- `callwaiting=no`: This is the standard feature that allows you to be on your phone line and be alerted to a second call trying to reach you. Call waiting is a nice feature, but your local phone carrier must carry the service. Otherwise, the second caller gets a busy signal before the call even hits your server.

- `threewaycalling=yes` **and** `transfer=yes`: We discuss both of these together because you can't have a `yes` on the first option without a `yes` on the second option. Three-way calling allows you to put the person that you are speaking to on hold and then dial another phone number to bridge into your conversation. To allow this to happen, you must enable Asterisk to transfer the first caller to hold while you dial the second person. Therefore, you need to have a `yes` on both options if you want this feature to work.

✔ echocancel=yes **and** echotraining=yes: Echo on a call can be anything from annoying to debilitating. Enabling the echo cancellation feature allows Asterisk to reduce the echo on the line. The echotraining option expedites the process by allowing Asterisk to send tones down the analog lines to determine the level of echo cancellation required.

✔ immediate=no: This option instructs the FXS port to respond to an outbound dialing attempt by either providing a dial tone or immediately executing a prescribed dialplan for the call. The feature is good if you have an Interactive Voice Response (IVR) system that handles all your incoming calls. If you want all your incoming calls to be sent to the default IVR system, choose immediate=yes. On the other hand, if you have individual Direct Inward Dial (DID) numbers set to ring into specific extensions and you are not using an IVR, your option is immediate=no.

Most people want the normal interaction of receiving a dial tone and inputting a phone number and not having a preassigned script execute a dialplan, so we recommend setting the option to no.

The configuration of immediate=yes requires the dialplan used for incoming calls to identify the generic s (starting) extension as the recipient of all inbound calls. If you are interested in the s extension, we discuss it in detail in Chapter 5.

Defining the channels

```
; define channels
context=incoming
signaling=fxs_ks
channel => 1

context=internal
signaling=fxo_ks
channel => 2
```

The last section in the zapata.conf file identifies the specifics of the two channels that you have configured. They list the context to which you've attributed them. The last two lines of each section simply identify the channel and the specific signaling. The context is an important aspect of the code because it represents a specific group of extensions that share the same routing parameters.

You can also aggregate individual channels together into group delineations as well as by context:

```
context=default
switchtype=national
signaling=pri_cpe
group=0
channel=>1-23
group=1
channel=> 25-47
```

In this instance, the `context/switchtype/signaling` are the same for both sets of channels. However, we have established two separate groups: The `group=0` command consists of channels 1–23, and `group=1` includes channels 25–47. We discuss the finer points of dialplans — including contexts — in Chapter 5.

Bringing in VoIP

The greatest benefit and flexibility of Asterisk is personified in its ability to handle VoIP calls. The protocol used by Asterisk is SIP. Many protocols are available to send and receive VoIP calls, but SIP is the easiest one to use and troubleshoot, as well as being the most efficient option. SIP doesn't handle the entire call, only the overhead. It establishes the call and does some minor housekeeping, but it doesn't send the actual voice portion of the call. The Real-Time Transfer Protocol (RTP) handles the voice portion. If you are feeling a bit rusty on your VoIP knowledge, check out Appendix B.

Asterisk uses the User Datagram Protocol (UDP) to transmit the SIP information. UDP doesn't resend information; it simply packs it up and sends it out in sequence. For that reason, UDP is standard for SIP transmissions.

SIP uses the standard communication port of 5060 for transmission. RTP is relegated to ports 10000 to 20000.

Getting to know the sip.conf file

The configuration of your standard SIP connection in the standard `sip.conf` file is as follows:

```
[general]
context=default
srvlookup=yes

[brady]
type=friend
secret=welcome
qualify=yes
nat=no
host=dynamic
canreinvite=yes
context=internal
```

The [general] section handles the default configuration for all SIP entries. The protocol is standardized, so this just takes care of the basics for handling all SIP calls. The information listed directs all SIP calls to the default context of [brady].

The srvlookup=yes is a parameter that is dictated by the VoIP carrier you are using. srvlookup (or service lookup) is the VoIP equivalent of the more common Domain Name Service (DNS) used in the IP world to translate domain names (such as www.google.com) into IP addresses (such as 64.233.187.99). Enabling srvlookup allows you to specify a domain name instead of an IP address; Asterisk then looks up the address.

In the following sections, we break down the rest of the sip.conf file.

Classifying friends, peers, and users

```
[brady]
type=friend
```

If you want a device to make only outgoing calls, you tag it as a *peer*. Conversely, to allow your device to only receive VoIP calls, you tag it as a *user*. *Friends* are both users and peers, endowed with the ability to both send and receive VoIP calls.

In our example, brady is the name of the device and is listed as a friend. This friend classification enables the brady device to send and receive VoIP calls.

Protecting yourself with secrets

```
secret=welcome
```

The secret section of the code identifies the password used to authenticate calls to and from the device by requiring the password. The secret section authenticates and registers remote devices that connect to your Asterisk, possibly a standard VoIP phone or a VoIP softphone.

In our example, welcome is the password.

Qualifying calls

```
qualify=yes
```

Instead of just randomly sending calls to a remote server or SIP phone, we recommend qualifying it first. Qualifying involves sending packets periodically to your remote devices so that your Asterisk server can confirm they are active. It is more efficient to check the pulse on your remote devices than to waste time sending an INVITE message and waiting for the attempt to time out, and then reinviting the call to voice mail or a default extension.

The duration for which a device remains qualified is up to you. We recommend that you don't qualify your remote devices for more than 60000 milliseconds, which is 1 minute.

NATing or not NATing?

```
nat=no
```

Network Address Translation (NAT) is a common feature of firewalls and routers. It translates IP addresses within your company's intranet into public IP addresses that are seen by the rest of the world. Set `nat=no` if you have a public IP address or `nat=yes` if you are behind a firewall.

We are speaking about device parameters. If your Asterisk is behind a NAT firewall, set `nat=yes` on the SIP server into which you are connecting. The `externip` parameter in the general configuration section of Asterisk helps overcome NAT issues by sending a specific IP address in the SIP message from your Asterisk server, as follows:

```
externip=200.201.202.203
```

Specify the IP subnet/network for local SIP devices located with your Asterisk server behind your firewall using the `localnet` parameter for each subnet/network, as follows:

```
localnet=192.168.0.0/255.255.0.0
```

Hosting on a static IP

```
host=dynamic
```

If you're using Asterisk for your office or as the platform for your business, you probably have a dedicated Internet circuit and a static IP address. You need at lease one static IP address so that external devices can speak to your Asterisk. This is regardless of whether you use DNS lookup (using Service Lookup records). If you're planning to run a public Asterisk server on the Internet, you must have a public IP address, either static or dynamic.

Dynamic IP addresses can cause connection problems. If your IP address changes (which is the nature of a dynamic IP address), you immediately lose connectivity from anyone who was connected to your server from your old IP address. Working with a dynamic IP address requires you to have a mechanism established to notify all devices that are or may be connected to your server when your IP address changes.

To avoid these problems with a dynamic IP address, always use static addresses when hosting from a public server. The remote devices register and authenticate with your server. This process is reliable and assures you of a safe connection.

Reinviting calls

```
canreinvite=yes
```

Technically, reinviting happens when an additional INVITE message is sent on a VoIP call after the initial INVITE. One of the best things about VoIP is its ability for the overhead of the call to connect directly between two locations, while the voice portion of the call in the RTP is sent to a completely different third destination. The RTP could be sent to voice mail, forwarded to a cellphone, or used to save the Internet bandwidth on your Asterisk.

In our case, we have the brady extension reinvite and forward to wherever it needs to go. The canreinvite parameter in the device configuration is specifically set up to reinvite packets to the terminating device after a call is connected. Only the voice portion of the call is redirected to the new endpoint, bypassing the hardware that allows the reinvites.

Reinviting is great for conserving bandwidth, because you aren't using 128K to receive the voice portion of the call (using 64K) and then sending it back out (using another 64K), but it doesn't work well if you are using a NAT firewall. The process of converting a private IP address to a public IP address becomes problematic when reinviting. This is because the voice portion of the SIP call could be emanating from a port within your LAN, existing behind the firewall. If Asterisk identifies the voice portion of the call is coming not from port 12000 in the unprotected area outside your firewall, but instead from port 1300 behind the firewall, Asterisk views the transmission as rogue RTP and you can easily lose the outbound audio stream. You will quickly realize that a problem exists when your calls have *one-way audio* (meaning that only one person on the call can hear the other).

If your Asterisk is functioning as your business platform, with a VoIP provider sending you calls on one side and the RTP being handled by Asterisk and sent out your same Internet connection to your client's server, you are using a lot of bandwidth. If you simply keep the SIP information coming through your Asterisk, but reinvite the RTP directly to your client's server, you free all the bandwidth and provide a better connection for the RTP. Every time the server handles the RTP, latency can be added, or call quality- or call-completion problems can be experienced. The fewer hands in the pot, the better.

Ending with context

```
context=internal
```

Context refers to the parameters for a specific group of extensions within Asterisk. See Chapter 5 for all the gory details on context.

Configuring an SIP user

SIP users are VoIP devices that call into your Asterisk server. Depending on the type of device you are using to connect into your Asterisk, you may have to configure all or only some of the following configuration options:

- ✔ `context`: The dialplan context for calls from this device.

- ✔ `permit`: IP addresses that this device can register from.

- ✔ `deny`: IP addresses that this device cannot register from.

- ✔ `auth`: The authentication line for this device. This may be a simple authentication, or it may also include the `secret` parameter that sets a secret password or an `md5secret` if you prefer your passwords to not use plain text.

- ✔ `dtmfmode`: Touch tones — what the telecom world calls Dual Tone Multi-Frequency (DTMF) tones — have two flavors. You can have either in-band or out-of-band RFC 2833.

- ✔ `canreinvite`: This option allows the device to reinvite the voice portion of the call to another device downstream.

- ✔ `nat`: If your SIP device is behind a firewall, you are NATing. If you aren't, well, you aren't NATing.

- ✔ `callgroup` and `pickupgroup`: This is a special parameter used in a specific configuration that you probably won't bump into.

- ✔ `language`: This is the language you will be speaking; it helps with the voice prompts from Asterisk.

- ✔ `allow`: This is where you list the allowed coder/decoders (*codecs*) for this device. We cover codecs in detail in Chapter 9.

- ✔ `disallow`: This is where you list the disallowed codecs for this device. For example, you may not want to use a certain codec with compression.

- ✔ `insecure`: This parameter identifies how much you trust this device explicitly.

- ✔ `callerid`: This is the outbound caller ID that you show for this device.

- ✔ `accountcode`: This is a billing code that is placed into the overhead to track the call in Asterisk for billing.

- ✔ `restrictcid`: This option hides the outbound caller ID.

Configuring an SIP peer

An *SIP peer* is a VoIP device using the SIP protocol that your Asterisk dials out to. When you are configuring your SIP peer, the configuration options are similar to those of an SIP user (see the preceding section).

In addition to the SIP options, you also have the following options for an SIP peer:

- `mailbox`: This represents the mailbox assigned to this device.
- `username`: This identifies the username the device registers with.
- `fromdomain` **and** `fromuser`: These are required only in special cases and it's noted in the device you are configuring (this is for carriers that are not specific to devices).
- `host`: You need the IP address or a dynamic IP when the device registers with your server.
- `port`: You list the port for the device if it is anything other than the standard port of 5060. This isn't used if you have a dynamic IP. In that case, you use the `defaultip` setting.
- `defaultip`: If you set `host=dynamic`, this is the default IP address prior to registration.
- `qualify`: Determines whether you need to qualify this device.
- `rtptimeout`: Establishes the maximum number of seconds Asterisk waits for RTP information before considering the call disconnected.
- `rtpholdtimeout`: Identifies the maximum number of seconds Asterisk waits for a call on hold before the call is considered disconnected.

Understanding InterAsterisk eXchange (IAX) Connections

InterAsterisk eXchange (IAX) is a transmission protocol that is similar to the VoIP protocol SIP. It isn't mandatory for the transmission of calls to or from the Asterisk servers; it's simply another option. Hardphones and softphones exist for IAX, made by Digium and a variety of other manufacturers. If you don't have IAX phones, you can still connect to VoIP or regular analog phones.

The IAX protocol uses port 4569 for the transfer of both the overhead of the call and the voice portion of the call. In this respect, it acts more like an analog or digital call than a VoIP call, even though it is sent via a port. Some people like IAX because it can aggregate all voice traffic to a single destination into a single stream in a process called *trunking*. Trunking helps to maximize bandwidth because it eliminates the need for individual streams of overhead to maintain the integrity of each call.

Setting up an inbound IAX connection

Inbound IAX connections are set up by modifying the iax.conf file located in the /etc/asterisk directory. Execute the make samples command to find this file:

```
[general]
bandwidth=low
disallow=lpc10
jitterbuffer=no
forcejitterbuffer=no
tos=lowdelay
autokill=yes

register => marko:secretpass@tormenta.linux-support.net

[brady]
host=dynamic
secret=mysecret
mailbox=1234
defaultip=200.201.202.203
callerid=brady Kirby

[iaxfwd]
type=user
context=incoming
auth=rsa
inkeys=freeworlddialup
```

The iax.conf file isn't that complex. The [general] section of the code covers the environment of all calls on the IAX device. Each of the options here must allow a variety of environments and end devices. The system must be able to accommodate the remote IAX softphones as well as hardphones wired into the network. This example portrays a device called brady that is wired into the local-area network (LAN) and is logging in to the server.

Setting up your security

```
auth=rsa
inkeys=freeworlddialup
```

These two lines of code establish the type of security used for the transmission and the filename used: `auth=` identifies the method of authentication used on your transmission. Options for `auth=` are as follows:

- ✔ `plaintext`: This option has plaintext access to the files used for encrypting and de-encrypting the authentication.

- ✔ `md5`: Like `plaintext`, this option has plaintext access to the files used for encrypting and de-encrypting the authentication, but it is also pre-encrypted.

- ✔ `inkeys`: The most secure method of authentication is `rsa`, which uses two files for authentication that are referred to as *public*(or *inkeys*) and *private* (or *outkeys*) keys. The `inkeys=` line of code identifies the name of the file on your server used by the remote device to authenticate your connection. The outkey is the file within your hardware that you don't share with the outside world and use to decode the received authentication.

Assigning your bandwidth

```
bandwidth=low
```

Just like the option used when you download video files, you can choose low or high bandwidth.

Disallowing audio and video

```
disallow=lpc10
```

The `disallow=` option is specifically used to deny the use of specific compression methods on audio and video transmissions. Chapter 9 covers how to choose your compression method.

Avoiding the jitters

```
jitterbuffer=no
forcejitterbuffer=no
```

All real-time voice transmissions love consistency. They work best when the packets for each call are sent in a constant stream with a uniform time between each packet. A delay in the transmission of the packets, causing them to arrive more slowly than expected, is called *latency*. Latency isn't horrible; a bit of it is actually expected as calls are converted from one protocol to another (for example, IAX to analog). Problems arise when the latency is excessive or when the duration of latency varies considerably. This fluctuation in latency is called *jitter*, and the `jitterbuffer` and `forcejitterbuffer` options help to reduce jitter.

We suggest that you configure these options to =no because the process of avoiding jitter by storing packets within your server, with the intention of sending them out in a uniform cadence, actually causes more latency. (This is because you are holding onto the packets instead of sending them directly, which is the definition of latency.) Only if you are functioning in a network that experiences a lot of latency should you configure these options to =yes.

Enhancing your type of service desired

```
tos=lowdelay
```

The *type of service (ToS)* option allows you to decide the performance option on your connection. The `tos` values mark the packets being sent to allow them to be handled with greater care, priority, or efficiency. The available `tos` options are as follows:

- `lowdelay`: Minimize delay.
- `throughput`: Maximize throughput.
- `reliability`: Maximize reliability.
- `mincost`: Minimize cost.
- `none`: No markers are placed on the transmissions, and no special handling is performed.

The tagging done by this option only works if the router that is passing the data understands and uses the tags. This is not a common feature for transmissions processed over the public Internet because anyone can tag a packet. The only way to use and respect the tags is to use them within the controlled environment of your LAN, providing priority for calls as they leave your network.

Killing registration attempts

```
autokill=yes
```

The `autokill` option pertains to the system's parameters to cease an attempted registration of a remote device if an acknowledgment is not received from it in 2000 milliseconds. The available options are =yes, =no, and =*XXXX*, where *XXXX* is a specific number of milliseconds you want to wait before killing the registration attempt.

Registering the call

```
register => marko:secretpass@tormenta.linux-support.net
```

Asterisk needs to register every call before the call is allowed to officially enter the server and be processed to an extension on the LAN. Registering is very important because allowing your Asterisk to process calls from unknown points of origin exposes you to potential fraud.

In our example, we're sending a registration to the server `tormenta.linux-support.net` with the credentials of `username=marko` and `password=secretpass`.

Classifying the call

```
[brady]
host=dynamic
secret=mysecret
mailbox=1234
defaultip=200.201.202.203
callerid=brady Kirby

[iaxfwd]
type=user
context=incoming
```

This last section of the code classifies the call by the context (in our case, it is an `[incoming]` context) and specifies the type of call you're expecting as well as the network from which it originates. Because you are receiving calls, you should anticipate the call to originate from a remote phone on a network (in our case, the remote phone is called `brady`). This classifies the type of connection as being configured for a user instead of a peer, just like you would see in an outbound call.

Designing outbound IAX connections

Setting up an IAX channel is the first part in sending and receiving calls from it. Specifying the device as a peer assigns the `[incoming]` context to it. If no peer exists, the device is assigned to a default context that handles incoming calls from a variety of contexts, such as ZAP, IAX, or SIP.

You can attribute outbound devices to a default `[outbound]` context based on the device they terminate into. The extensions for outbound extensions list the IP address or DNS hostname of the server. The standard configuration in a dialplan looks like this:

```
[outboundcalls]

exten => 2567142943,1,Dial(IAX/fwdoutbound/2567142943)
exten => 2563192010,1,Dial(IAX/207.111.170.18/2563192010)
```

Receiving Help with Debugging

Debugging is used to show the messages passed on the specific channel that's experiencing challenges and which you are trying to analyze. The Asterisk code is loaded with messages and shreds of information that the developers thought would be helpful. You can use this info in troubleshooting an improper software implementation or a device that is processing a call.

Execute debug settings from the main Asterisk command-line interface (CLI). You reach this by executing the following command from the Linux command prompt:

```
asterisk - r
```

After you reach the Asterisk CLI, enable the debug options for the connections by executing the commands:

```
iax debug -vvvv
sip debug -vvvv
```

Each command has specific options to debug a device, and the number of vs you add to the end of the command correlates to your level of *verbosity* (how much information you want). The more vs you add, the more information you get.

Troubleshooting applications, channels, and modules is done with greater precision by using commands in the following syntax:

```
modulename debug <parameters>
```

In this instance, the parameters apply to the specific application, channel, or module you are analyzing. Debugging a specific SIP module at the IP address of 207.111.170.18 uses the following code:

```
Sip debug IP 207.111.170.18
```

You can use this syntax for all your debugging needs. After you finish troubleshooting, you deactivate the debugging with the following syntax:

```
modulename no debug
```

Discontinuing the debugging on the SIP port with the IP address of 207.111.170.18 uses the code:

```
SIP no debug
```

Part II
Using Dialplans — the Building Blocks of Asterisk

The 5th Wave   By Rich Tennant

"...for technical support, press 7; for product information, press 8; if you're bored and just want to argue with someone, press 9..."

That's me...

In this part . . .

You discover the things that make Asterisk tick. Chapters 5 and 6 cover the Asterisk dialplan, the essential building blocks of Asterisk. You find out the proper syntax and structure required to write the dialplan (Chapter 5). You can enhance your dialplan with variables, macros, expressions, operators, and functions (Chapter 6). We also show you how to enable conference calls, voice mailboxes, call queuing, and music on hold.

Asterisk has a few nice features that allow you to view channels, monitor calls, and record information (Chapter 7). We walk you through all the useful nuances of these features and the Asterisk Database. In Chapter 8, we show you how to configure AsteriskNOW through its GUI interface. In Chapter 9, we give you the skinny on all the VoIP codecs.

Chapter 5

Comprehending Dialplan Syntax

· ·

In This Chapter

▶ Comprehending dialplan syntax

▶ Adding extensions for the inevitable growth

▶ Building dialplans for incoming calls

▶ Designing internal dialplans

▶ Utilizing variables in dialplans

▶ Using pattern matching for cost savings

▶ Building outbound dialplans

· ·

*I*f you imagine Asterisk as a creature, you can think of the server itself as the body and the Zaptel cards, IP interfaces, network interface cards (NICs), and InterAsterisk eXchange (IAX) connections as its arms, legs, hands, and feet. The Asterisk kernel is the brain, and the dialplans function as the central nervous system linking all the components together and allowing them to move and perform tasks.

Dialplans pass information and calls through the network. Your Asterisk server can make an outbound call over the Zaptel card on port 1 provisioned as a Foreign Exchange Office (FXO) interface, instead of just knowing that card exists. You can set up your Asterisk server to handle each type of call — whether an inbound, outbound, or internal call — differently, but all code for any application uses the same naming convention in the structure of the extensions.conf file.

Starting with a Basic Dialplan

Dialplans define the exact parameters of how a call is processed. When someone calls you, do you want the call to go to your extension first and then your voice mail? Or, do you want the caller to receive a recording asking him to enter an extension first? You can also establish an order in which calls reach you. You can build a dialplan that attempts to reach you on your office line, your cellphone, and the VoIP softphone on your computer, and then send the call to your voice mail . . . now that's being connected!

You build the dialplans in the `extensions.conf` file that is usually located in the following directory:

```
/etc/asterisk
```

The `extensions.conf` file is one long string of dialplans, separated only by the individual contexts within brackets. All configurations beneath one context and above another apply to the upper context.

This is an example of a simple dialplan:

```
[incoming]

exten => 2565551212,1,Answer()
```

This simple dialplan doesn't look like much now, but from these humble beginnings, great things are made.

Figure 5-1 identifies the four elements of a basic dialplan. Each element of the dialplan acts as shorthand for a set of rules, processes, or locations that are applied to all calls that fall within the dialplan parameters. The dialplan in Figure 5-1 is for an incoming call that is being directed to phone number 9495551212. This scheme doesn't play a greeting to the caller, nor does it offer voice mail or a list of extensions; it simply sends the caller to the extension for phone number 949-555-1212, where the caller remains until someone picks up the line or he hangs up the phone.

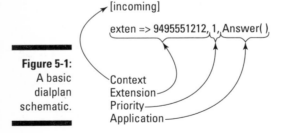

Figure 5-1:
A basic dialplan schematic.

We discuss these four elements — the context, extension, priority, and application — in the next sections.

Understanding the context

```
[incoming]

exten =>9495551212,1,Answer()
```

The *context* is the first line of code and it correlates to the context you defined when configuring the individual VoIP, analog, digital, and/or IAX devices. (If you're wondering how to program the context to these devices, check out Chapter 4.)

When a call comes in to the server, Asterisk looks for the device sending the call (SIP/0001, for instance) for registration. If Asterisk can't find the registration, it looks for a default context for the channel type (VoIP, analog, or IAX connection, for example).

You should group devices with the same functions and that belong to the same financial entity into contexts for ease of programming. The contexts we cover in this chapter are as follows:

✔ [incoming]: Use this context to configure Asterisk for inbound calls.

✔ [internal]: Use this context to configure Asterisk for calls that are transferred between two endpoints within the Asterisk network.

✔ [outgoing]: This context is for outbound calls through analog, digital, VoIP, or wireless connections.

✔ [globals]: This is a special context in which you build dialplan universals that you apply to all contexts. Your home area code would be listed here, for example, and used as a telecom point of reference for all local calls.

These contexts are actually configuration contexts assigned to devices connecting to the Asterisk system. The device parameters and abilities are linked to the dialplan, which allows you to process the incoming and outgoing calls from the device. The [globals] context is the only special context because the relationships established in this context are used in every other context in your dialplan. These relationships are generally coding expedients whereby a common name is used to replace a complex device ID, such as

```
TRUNK=SIP/111.222.333.444/
```

This programming shorthand, called a *variable,* allows you to write the word TRUNK instead of typing the VoIP port and IP address of SIP/111.222.333.444.

You don't have to use all these specific context names. Only the [globals] context is preset in Asterisk. You can use any naming convention you want for the other contexts. Context names can include letters, numbers, and hyphens. You can make your life easier if you do name your dialplans something descriptive. Context names such as [inbound], [outbound], [longdistance], [local], [international], and [extensiononly] are easier to work with, and you can more easily remember their purpose.

Okay, you can also use underscores in context names, but they can be problematic. Underscores are also used in other aspects of Asterisk, and mistyping a context with an underscore can complicate your dialplan. We recommend not using them for contexts; stick with hyphens instead.

If you're using Asterisk for internal company use as well as for external customers, create context names that reflect the different entities. For instance, you can split the [incoming] context into the following:

```
[incoming-mine]
[incoming-customer]
```

Splitting contexts reduces confusion and allows you to handle internal calls differently. If you dial 0 in your office, you probably want to speak to your receptionist. If your customer has the option of dialing 0, you can have him sent to your customer service queue instead.

Asterisk does have one context that isn't really a context as we understand them in the dialplan. The [general] context is located in the extensions. conf file, but is actually a general holding place for information in the extensions.conf file and not a traditional context like [incoming] or [outgoing]. It holds information that pertains to all contexts, like the autofallthrough option that acts like a security net to handle calls that may have inadvertently have no real destination due to a gap in your dialplan programming.

Identifying extensions

```
[incoming]

exten =>9495551212,1,Answer()
```

In the most traditional sense, *extensions* refer to the number you dial to call someone within your company. Asterisk takes extensions to a whole new level. It views not only the physical phone on a desk or channel driver (SIP/ZAP, IAX) as an extension, but it also provides intelligence to the extension allowing it to play a message while sending the call to various places, in this order:

1. The phone on your desk.

2. Your cellphone, if you don't answer your desk phone in a preset number of seconds.

3. Your voice mail, if you don't answer your cellphone.

The new era of VoIP also expands how you identify extensions. You are no longer restricted to simply using numbers to identify Asterisk extensions. You can use e-mail addresses as valid extension names if you build them into the system, as well as employee names (for use in dial-by-name directories).

The extensions you build can be as complex and dynamic as you need. The automated systems that prompt you for your essential and nonessential information when you call customer service are one example of the intelligence you can build into an extension, or a set of extensions. Every response the caller keys into the system either sends her to another extension within the system in a different context, or simply further along in the dialplan. As long as you remember that an extension can be a virtual destination within a device used to qualify responses of the caller, and not just a phone on a desk, you are ready to unlock the potential of Asterisk extensions.

Devices are physical hardware and can contain and facilitate extensions, but aren't extensions themselves. The only way you can reach a device is through an extension, which must be dialed to route a call to a specific device. The rub is the fact that extensions are not defined on a one-for-one basis with devices.

Extensions can have a multitude of possible destinations, of which only some are devices. Samples of nondevices into which an extension can terminate are as follows:

- ✔ Applications
- ✔ Direct voice mail servers
- ✔ Calling card applications
- ✔ Conference rooms
- ✔ Fax servers

If you want to call the physical analog phone in your customer service department from the VoIP softphone on your desk, you simply dial the Asterisk internal extension that directs your call to a device connected to the Zaptel card, and the remote phone rings. If you want to call Papua, New Guinea, over your outbound VoIP carrier, your Asterisk actually connects your call to an extension that terminates in an outbound device, such as SIP/1.

Extensions are identified in the `extensions.conf` file with the following line of code:

```
exten => 133,1,Answer()
```

The `exten =>` part identifies the extension, and the number or letters after the `=>` indicate the specific extension you're defining. In this example, 133 is the extension.

Just like the context names, you can design most of your own extension names. The default extensions with Asterisk are as follows:

- ✔ `exten => s`: This extension is the start extension. It receives calls and plays an automated greeting prompt as long as it is configured in your device configuration.

- ✔ `exten => i`: This is the invalid extension. If you prompt a caller to input a 3-digit extension number and he presses three digits that aren't listed as a legitimate extension, he's sent to `exten=> i`.

- ✔ `exten => t`: This is the timeout extension. You set time limits on some extensions in your dialplan, and if the caller doesn't choose a prompt within your time limit, the call is sent to the `exten => t` extension.

- ✔ `exten => T`: This is an *absolute* timeout identifying the maximum duration Asterisk allows for an active phone call before it's sent to the `T` extension or hung up. Carriers use a feature like this to prevent calls that have failed to completely hang up (called a *hung call*) from remaining active in their switch.

- ✔ `exten => #`: This functions like a command more than an extension. It's normally used to hang up a call in a dialplan with many layers of recordings prompting you to "press 1" for this and "press 2" for that. These Interactive Voice Response (IVR) systems use the extension as an expedient to disconnect a call, in spite of it not being a traditional "predefined" extension destination.

These extensions are case-sensitive. Confusing a `t` extension with a `T` extension could result in calls being disconnected that simply can't find their extension. Pay attention when using them in your dialplan to avoid headaches later.

These extensions allow you to build a solid dialplan. Asterisk does have a safety net to prevent calls from being sent to limbo and left hanging. The `autofallthrough` option in the `[general]` section of the extension configuration prevents forgotten calls from lingering in the system.

The default extensions in the previous bulleted list are predefined destinations in Asterisk. You must still define them in your context so that Asterisk knows what you want it to do when someone must be routed to the extension. Do you want to hang up on people that exceed your timeout limit, or send them back to the main recorded greeting for a second chance? Failing to program for these little eventualities can cause the call to congest the system and automatically be hung up.

It is very important to identify and process calls that time out. Someone could call you on Friday at 5:00 p.m. on your toll-free number, connect to your phone system, and fail to hang up correctly (honest, it happens). In this case, you would have a billable phone call that lasts until Monday morning, when the person calling you realizes his phone line has no dial tone because it is still connected to your system.

Utilizing priorities

```
[incoming]

exten =>9495551212,1,Answer()
```

The second parameter of the extension code is the *priority;* it controls the sequence of events in the dialplan. Each priority contains a call to a single application in Asterisk. It can also forward information to that application, if necessary, to provide additional features to the call. The priorities increase in value as each step in the dialplan is executed. A dialplan with several steps would look like this:

```
exten => 133,1,EatPizza()
exten => 133,2,DrinkSoda()
exten => 133,3,WathchTV()
exten => 133,4,TakeNap()
```

This is, of course, a bogus Asterisk extension with non-existent applications, but it does demonstrate how the priority increases sequentially as progressive actions are taken in sequence.

If you don't want to remember the last priority you used in an extension, you can replace the priority number with n (which is referred to as an *unnumbered priority*). The n represents the previous priority plus 1. You write an unnumbered priority in a dialplan like the following:

```
exten => 133,1,EatPizza()
exten => 133,n,DrinkSoda()
exten => 133,n,WathchTV()
exten => 133,n,TakeNap()
```

Engaging an application

```
[incoming]

exten =>9495551212,1,Answer()
```

The *application* is the last element of code in the extension line. Even though it is only one element, it is actually constructed of two distinct sections: the application and the argument:

✔ **Application:** The application represents the action being taken at this step (priority) of the extension. In Figure 5-2, the application being used is `dial`. The first channel on the Zap card from extension 9495551212 is being dialed.

Many different applications are available with Asterisk; `answer`, `hangup`, `background`, `dial`, and `Goto` are the most important to know, so we discuss those in the following sections.

Figure 5-2:
An application schematic.

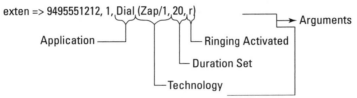

exten => 9495551212, 1, Dial (Zap/1, 20, r)

Application ──┘
Technology ──┘
Duration Set ──┘
Ringing Activated ──┘
Arguments ──┘

✔ **Argument:** The argument includes three unique parameters, separated by commas. In Figure 5-2 the arguments are as follows:

• **Parameter 1:** The syntax of the first parameter of the code specifies the channel technology/resource. Our example identifies the channel technology as ZAP, indicating the Zapata time-division multiplexing (TDM) (analog) card. The resource, listed after the backslash is 1, indicating port 1 as the channel device using this application. It could have just as easily been a device with a VoIP port, a digital card, or a Wi-Fi port.

• **Parameter 2:** The 20 identifies that you're requiring Asterisk to maintain this application for 20 seconds. If the dialed device answers the call before the 20 seconds are up, the call is bridged through it. If the dialed device doesn't answer in 20 seconds, the extension proceeds to the next priority. You can configure it to send the call to voice mail, return to the main operator for the system, or disconnect the call. Each of these available fallback destinations is an additional priority and is located in the current context, or in another context.

• **Parameter 3:** The r parameter allows the person originating the call to hear the phone ringing.

You can omit any of the parameters within the argument if you don't want to restrict the call or enable the ringing. An extension without a specified duration looks like this:

```
exten=> 9495551212,1,Dial(Zap/1,,r)
```

If you want to keep the 20-second time limit on the application, but don't want to the person making the call to hear a ringing tone, the code is as follows:

```
exten=> 9495551212,1,Dial(Zap/1,20)
```

Answering calls

The `answer()` application doesn't use arguments; because the parentheses are always blank, you can omit them if you want. The `answer()` application receives a call from the outside world and sends a connect signal back through the device that originated the call.

The parentheses in applications aren't essential if they don't hold any information. You can even replace the parentheses with a single comma between the application and the arguments. For example, you can write the application `Dial(Zap/1,,r)` as `Dial,Zap/1,,r`.

The Asterisk environment is very dynamic, so don't be linked to the idea that whenever we mention a call that it is coming from someplace beyond your network. A wealth of calls is flying through Asterisk networks from SIP phones to voice mail, or analog phones linking up on a teleconference. A whole world of traffic is rolling through your Asterisk phone; this traffic is referred to as *calls*.

The *connect* signal sent back to an incoming carrier happens regardless of the device that delivers the incoming call. The `answer()` application also receives inbound VoIP calls or internal calls into your Asterisk phone. The `answer()` application is only necessary when immediately connecting you to a system that plays a message, such as "Please enter the extension of the person you wish to dial," before connecting you to a second device.

The application doesn't play a recorded message or anything else, aside from establishing a connection on the call. The code for a call sent to a default start extension (s) and answered appears like this in a dialplan:

```
exten=> s,1,Answer()
```

Only use the `answer()` application when it is inside your IVR system (you know, the thing asking you for the extension you wish to dial). You don't have to use this application if you are using the `dial` application directly because it automatically answers the call when the destination channel picks up the call.

Hanging up calls

The *hangup* is the logical conclusion to every call. As such, this command doesn't appear with a priority of 1. Common courtesy dictates that you at least answer the phone before hanging up on someone. The code for a call to be answered and then hung up is as follows:

```
exten=> s,1,Answer()
exten=> s,2,Hangup()
```

Responding to touch tones

We have all called a large company and received the voice-mail prompt of "Please enter your party's 3-digit extension." You also notice that if you dial the 3-digit extension while the recording is playing, it stops the recording and immediately sends you to the extension you dialed. Asterisk has the same feature, and because it is waiting in the background for you to dial the extension, the application is called background(). It looks like this in the dialplan:

```
exten=> s,1,Answer()
exten=> s,2,Background(enter-ext-of-person)
```

Dialing the new-fashioned way

Every inbound call includes at least two segments. The [incoming] context receives the call into the Asterisk, but after it has arrived there, the call must still be sent to a desired channel device, voice mailbox, or external carrier. The dial() application allows you to forward the call to its next destination. The following example demonstrates a simple dialplan using dial():

```
[internal]
  exten => 0,1,Dial(Zap/1,20,r)
```

This code allows internal customers to dial 0 and receive the operator that is connected to the Zap/1 card.

Moving with a Goto

The Goto application allows you to do wonderful things because it allows you to send a call from one context to another. A simple dialplan that allows an incoming customer call to reach your internal operator looks like the following:

```
[internal]
  exten => 0,1,Dial(Zap/1,20,r)
[incoming-customers]
  exten => 7,1,Goto(internal,0,1)
```

Asterisk applications

If you want a complete list of the applications available on your Asterisk, you can type the `show applications` command in the Asterisk command-line interface (CLI). When you find an application you are interested in, you can read a detailed document on the application by querying Asterisk for it with the following command:

```
Show application name
```

Simply replace *name* with the name of the application you are interested in, and a document appears with all the application's pertinent information. You can also find the documentation by scanning the /docs/ directory of the Asterisk source code.

For a general list of applications available on Asterisk, check out the following Web site:

```
www.voip-info.org/wiki-
    Asterisk+-+documentation+
    of+application+commands
```

This simple bit of code shows how the call in the `[incoming-customers]` context, when pressing extension 7, is routed to the new `[internal]` context, extension 0 and priority 1. The `[internal]` context identifies where extension 0 is sent.

Processing Incoming Calls

After you set up an extension, you can move ahead with some standard applications and design a set of extensions and contexts to make them work. Consider the following context and extensions:

```
[incoming]

exten => s,1,Answer()
exten => s,2,Background(enter-ext-of-person)
exten => 133,1,Playback(vm-nobodyavail)
exten => 133,2,Hangup()
exten => i,1,Playback(pbx-invalid)
exten => i,2,Goto(incoming,s,1)
exten => t,1 Playback(vm-goodbye)
exten => t,2,Hangup()
```

This context is well designed, as long as extension 133 is your only active phone in the office. A call has the following three possible outcomes in the `[incoming]` context:

✔ The call is answered by Asterisk, and the caller inputs the number 133 and is sent to that extension. If the call is not answered, a recording states that the person is unavailable and the call is disconnected.

✔ The call is answered by Asterisk, and the caller inputs an invalid extension and is sent back to the [incoming] context, extension s, first priority.

✔ The call is answered by Asterisk, and the caller doesn't input an extension. Asterisk plays a good-bye recording, and the call is disconnected.

The general timeout for a priority is roughly five seconds. If you require more time than that, use the waitexten application to specify how long to wait before the call times out.

Building on the dialplan is simple; just type more extensions after the first extension. Adding extension 137 to the dialplan looks like this:

```
[incoming]

exten => s,1,Answer()
exten => s,2,Background(enter-ext-of-person)
exten => s,3,WaitExten(10)
exten => 133,1,Playback(vm-nobodyavail)
exten => 133,2,Hangup()
exten => 137,1,Playback(vm-nobodyavail)
exten => 137,2,Hangup()
exten => i,1,Playback(pbx-invalid)
exten => i,2,Goto(incoming,s,1)
exten => t,1 Playback(vm-goodbye)
exten => t,2,Hangup()
```

You need to build an extension for every device in a context that receives a call. If you only build a context with two extensions, every other option is invalid and is sent to the i (invalid) extension.

Building Internal Options

So far in this chapter, we discuss the calls coming into your Asterisk system from the outside world, but you need to also build connections for internal users. Your salesman with a VoIP softphone in Prague would love to be able to call your office manager in Chicago by just dialing a 3-digit extension, so you have to build that capability into your Asterisk system. By creating an [internal] context and listing specific extensions, you allow the extensions to be connected so that someone can dial an extension you attribute. The code is as follows:

```
[incoming]

exten => s,1,Answer()
exten => s,2,Background(enter-ext-of-person)
exten => 133,1,Playback(vm-nobodyavail)
exten => 133,2,Hangup()
exten => 137,1,Playback(vm-nobodyavail)
exten => 137,2,Hangup()
exten => i,1,Playback(pbx-invalid)
exten => i,2,Goto(incoming,s,1)
exten => t,1 Playback(vm-goodbye)
exten => t,2,Hangup()

[internal]

exten => 207,1,Dial(SIP/Brady,20,r)
exten => 210,1,Dial(ZAP/1,20,r)
```

We've listed only two extensions in the [internal] context, so only SIP/ Brady can dial the 3-digit code of 210 to connect to the person at ZAP/1. Of course, the person at ZAP/1 can also dial SIP/Brady by simply dialing 207.

Using Variables

Just as you have s, i, and t extensions and the unnumbered priority n to make your life easier, other variables are available to Asterisk to make your programming life easier. At the most basic level, a *variable* is an easy-to-remember replacement for an element of code that you use continually. The most common variables you use are those that represent specific channel devices.

The following is an example of a variable:

```
Venus=ZAP/g0/4155551212
Mars=SIP/voip.atlasvoip.com/2055551212
Saturn=IAX2/voip.atlasvoip.com/4145551212
```

The variable Venus represents a channel linked to a port on a Zaptel card, Mars is a channel on a VoIP connection, and Saturn is an extension on an InterAsterisk eXchange (IAX) port. These variable names are much easier to remember and type than the specific information for the extensions.

This is the main benefit: If you need to reassign a new channel device to the Venus variable, you simply make the change in the variable, and all the connections you have made to Venus then reference the new channel device.

To reference the value of the variable in a line of code, you must represent it with a leading dollar sign ($) and the variable bound by braces. To use the Venus variable, you write the code like this:

```
exten => 133,1,Dial(${Venus},20,r)
```

The area to which the variable applies provides its name and the territory over which it's used. Three types of variables that help you with your dialplans are as follows:

- Global variables
- Channel variables
- Environmental variables

Exploiting globally

Global variables are listed under the [globals] context and are known to all contexts in your Asterisk. The following is a sample extensions.conf file with global variables:

```
[globals]

TRUNK=SIP/voip.atlasvoip.com
Venus=Zap/g0/4155551212
Mars=${TRUNK}/2055551212
Saturn=Sip/Brady

[incoming]

exten => s,1,Answer()
exten => s,2,Background(enter-ext-of-person)
exten => 133,1,Playback(vm-nobodyavail)
exten => 133,2,Hangup()
exten => 144,1,Dial(${Saturn})

[internal]

exten => 207,1,Dial(${Saturn},20,r)
exten => 210,1,Dial(ZAP/1,20,r)
exten => 2563192010,1,Dial(${TRUNK}/2563192010)
```

In this example, both the [incoming] and [internal] contexts have integrated the global variables.

Focusing on the individual call

Channel variables only interact at the individual call level. Channel variables are useful when you want to keep track of information on a call through each application. The ability to track the evolution of a call is most helpful when you advance a call from your Asterisk onto another extension or device.

Figure 5-3 displays the path taken by a call in the following dialplan:

```
[incoming]

exten => 777,1,Set(__forwardhelper=${EXTEN})
exten => 777,2,Dial(SIP/brady)

exten => 2565551212,1,Set(CDR(userfield)=${forwardhelper})
exten => 2565551212,2,Dial(Local/${EXTEN}@outbound)
[outbound]

exten => 2565551212,1,Dial(${TRUNK}/${EXTEN})
```

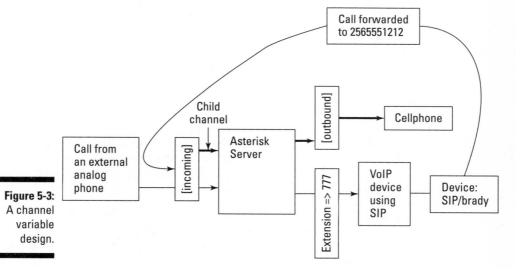

Figure 5-3: A channel variable design.

The diagram and the code look scary, but after the call is broken down, you can use it to make a template for other scenarios where you need the functionality.

The sequence of events in the dialplan is as follows:

1. An [incoming] context call is received by Asterisk and dials extension 777.

   The dialplan lists an extension 777 to receive the call. Before it is forwarded, you attribute the 777 value of the extension with ${EXTEN} to the call with the variable __forwardhelper=.

   The forwardhelper variable isn't a standardized Asterisk variable; you could have called it anything. This allows you to identify the extension dialed further down the dialplan by referencing ${forwardhelper}.

2. Extension 777 dials the VoIP device named SIP/brady.

   The VoIP phone at the SIP/brady extension has been programmed to forward all calls to a 256-555-1212 number.

3. The SIP/brady extension knows that the call must be routed through the Asterisk, so it forwards it back through the incoming context.

   The Call Detail Record (CDR) field on the call is now branded with 777 in the UserField. This is because you used the SET variable in the UserField portion of the CDR application to the value of forwardhelper, which is 777. By tagging the call with the extension, you can now identify this call in the CDR; this includes the record of its start time, stop time, and dialed number as well as the newly branded UserField.

4. The call is sent out via a local extension to the [outbound] context and creates a child channel, which is a second call that originates from the Asterisk.

5. The call to the device in the [outbound] context is terminated.

Variables prefixed with a double underscore (__) apply to all child channels on a call. If the call needed to be rescinded to the Asterisk again and forwarded out, the double underscore before the forwardhelper variable ensures that the 777 extension and CDR UserField are held throughout the life of the call. If the forwardhelper variable was prefixed with a single underscore, Asterisk would only apply the tag to the first child channel used.

In the previous sequence, we used the call forwarding on the SIP/brady extension to direct the call back to the [incoming] context, generating a second CDR. We could have built the same result into the dialplan using the following Goto application:

```
goto(outbound,2565551212,1)
```

The Goto application would result in only a single CDR for the call and the same forwarding. The danger in this method is that the CDR created by the Goto application gives the appearance that your system has been compromised. When you review your CDR to check for anomalies, you expect to see only calls from internal extensions making calls to destinations outside your network. The Goto application generates a CDR that lists the origination phone number calling into your network and terminating at a destination outside your network. Until you determine how this happened, the first logical conclusion is that an unauthorized entity has hacked into your system to make phone calls at your expense. Breaking the call into two CDRs duplicates the accounting for the call but prevents you from worrying about your network security.

The Local channel type in the dial application for the number being forwarded to specifies that the call is creating a second channel (sometimes called *forking into a second channel*). This second channel (sometimes called a *child channel*) begins in the [outbound] context at extension ${EXTEN}. The ${EXTEN} variable is another channel variable built into Asterisk for the current extension.

Using local extensions

Local extensions are commonly used in the "find-me-follow-me" services provided by many VoIP carriers. This scenario is also helpful if you are simply using Asterisk as your phone system and trying to reach people at their desk phone before retrying the call by sending it to their cellphone.

The following dialplan uses local extensions in sequence, allowing one call to attempt three different extensions:

```
exten => 123,1,Dial(Local/234@${CONTEXT})
exten => 234,1,Dial(Local,345@${CONTEXT})
exten => 345,1,Dial(SIP/${TRUNK}/2565551212)
```

Dialing the 123 extension creates a secondary channel starting at extension 234 in the same context. This extension generates a third channel at extension 345 in the same context, finally calling 2565551212 using the SIP technology and using the ${TRUNK} proxy server to pass the call through.

Dialing a local channel is different than dialing an extension that terminates at a VoIP or analog device. The codes do look similar:

```
dial(Local/XXXXXXX)
```

versus

```
dial(SIP/XXXXXX)
```

The dial(Local/XXXXXXX) code isn't VoIP or analog, but rather is a local channel that resides on the Asterisk server. Dialing the local channel creates a secondary channel, similar to the outgoing channel of a normal call. The difference is that the connection to the local channel creates a second record for the call. The initial connection into the Asterisk is identified by a destination channel of `Local/XXXXXXXX,1`, while the second portion of the call to the local channel (sometimes called a *subchannel*) has a source channel of `Local/XXXXXXXX,2` but links back to the first call detail record. If you are only using Asterisk for your internal phone system, this isn't an issue. If you are reporting off of this information, and using your Asterisk server as a carrier platform, you must link the calls together or risk losing a portion of the call, and also losing the revenue for that call.

Concerning yourself with the environment

Environment variables pertain to the Linux environment and are designed for use by programmers that have too much free time. These variables are helpful if you want to integrate some code into a dialplan to gather information about the processor speed and available disk space and play it back to you through the IVR, but it really doesn't help you with your dialplans.

This code shows how to use the environmental variables:

```
${ENV(varname)}
```

Adding Intelligence to Your Dialplan

Routing calls from an `[incoming]` context to an `[internal]` or `[outbound]` context based upon the number dialed may be fun, but it's limiting. Asterisk is capable of so much more. Building logic into your dialplan with special code and useful applications enables you to evolve it beyond the normal phone system.

Ignoring leading digits

Some people have to dial 9 for an outside line, and some long-distance carriers don't require you to dial a leading 1 for domestic calls. Even if you instinctively dial 9 for an outside line at your business, you don't have to retrain yourself with Asterisk. By adding the 9 to the beginning of the pattern-matching code, you can route the call through the Asterisk and then strip it off before sending the call to the outside world. No long-distance phone switch at any network accepts a 9 in front of a valid phone number. If you leave it in the call stream, every call will fail.

Luckily, Asterisk comes ready to help. You can use the `${Exten: }` syntax to strip off as many leading digits as you want. So where your dialplan requires an initial 9, you only need to strip off one digit; the command looks like this:

```
${Exten:1}
```

When you see this in a dialplan, the line of code appears as follows:

```
exten => _9NXXXXXX,1,Dial(${OUTBOUNDTRUNK}/${Exten:1})
```

To strip off more digits, simply change the number after the colon in the expression. Stripping three digits from the beginning of a dialed number requires the following syntax:

```
${Exten:3}
```

Cashing in with account codes

The `accountcode` variable allows you to link a unique code, traditionally a 2- to 7-digit number, to an individual call. Of course, you can only take advantage of this variable with devices that inherently recognize account codes. You can tag an account code to a call at any point within a dialplan. The end goal is to generate a report based on the account code, which allows you to group calls based on the account codes.

For example, you can track the success of a regional marketing plan by tagging sales department calls. If you're launching a sales campaign in Milwaukee, Wisconsin, it would be helpful to know the number of calls you receive from that area for the next 30 days.

The `set()` command is the most common way to establish an account code for an extension. The standard syntax for all `set()` commands is as follows:

```
set(variable=value)
```

The following line sets the account code variable for extension 333 to user `BKirby`:

```
exten=> 333,1,set(accountcode=bkirby)
```

Asterisk doesn't have a preset list of reports available, so enlist the skills of an accomplished programmer and build a program yourself. You can save time by using a program that someone else built, but you sacrifice response time when adding or modifying features. We recommend investing in the manpower required to build the software in-house. The intricacies of your dialplan may be specific to your company, your industry, or the needs of your customers, and the nuances of your required system may not communicate efficiently to a third-party software provider. If you are using Asterisk for internal use only on a straightforward application, prebuilt software may be all you need. However, if your business depends on the features of the reports and the visibility they provide, write the code yourself.

Replacing caller ID

Everyone loves their caller ID. Caller ID lets you make an informed decision on whether you really want to answer an incoming call or not. Asterisk has a variable that allows you to change the caller ID on a call. The RDNIS (*ReDirecting Number Inward Supervision*) variable does the same thing as the similarly named CALLERID(rdnis) application, and allows you to replace the true caller ID that originated a call with another phone number. If you are functioning as a carrier, it assists you in billing your customers because it allows you to attribute calls effectively.

For example, if you call your boss, who has an analog phone attached to your Asterisk network, and she is unavailable, you can have the call forwarded to her cellphone. Your boss's extension number is the originating phone number for the second leg of the call.

Manipulating the caller ID on a call to hide your identity or manipulate the rates charged by your carrier can land you in hot water with your carrier or the Federal Communications Commission. As a general rule, you need to leave the Caller ID field alone.

More information is available on RDNIS at www.voip-info.org/wiki-RDNIS.

Dialing to the Outside World

The following two steps are required to enable you to dial outside your network:

1. Establish an outbound trunk on which to dial.

2. Link that outbound trunk to a context.

Back in the 1940s, you might have picked up your phone and received not a dial tone, but an operator asking you for the number you wanted to be connected to. You might have told her, "Connect me to Mayberry 30924," and she would connect your call. Well, dialing has evolved so much that you can now place a call — a VoIP call that is — to SIP/Mayberry309@smalltown.com.

Asterisk allows you to build a Mayberry309 extension in your default context that routes your call. Extensions in Asterisk can include characters such as the plus sign (+) and the at sign (@) as well as numbers and letters.

A standard outbound dialplan looks like this:

```
[globals]
TRUNK=SIP/voip.atlasvoip.com
TRUNKLD=SIP/voip.ldtermination.com

[outbound]
exten => _91NXXNXXXXXX,1,Dial(${TRUNKLD}/${EXTEN:1})
exten => _91NXXNXXXXXX,2,Hangup
exten => _9NXXXXXX,1,Dial(${TRUNK}/${EXTEN:1})
exten => _9NXXXXXX,2,Hangup
exten => _9011.,1,Playback(international-calls-not-
            allowed)
exten => _9011.,2,Congestion
```

When dialing out to the Public Switched Telephone Network (PSTN), you have no fallback — you don't have voice mail when you make an outbound call to the PSTN for the PSTN number that you called. Congestion immediately terminates the current call with all certainty that the call is a failed call, not busy or a hang up call.

The `congestion()` application terminates the call and plays the Asterisk sound file of a fast busy signal in the event that your outbound call fails. This is Asterisk's only option when a call hits a line that is busy, or the call is answered and hung up normally. If the number you've dialed is disconnected or no longer in service, if the area code has changed, or if you simply misdialed, you don't hear the standard recordings that you do through a direct analog phone.

The exact recording you receive on a failed call can be very beneficial to understanding why the call failed. Take a look at Chapter 13 (the basics of telephony troubleshooting), Chapter 14 (analog, VoIP, and IAX calls), and Chapter 15 (digital calls) to find out the source of failed calls.

Understanding pattern matching

Before you can configure your Asterisk to dial out to the rest of the world, you need a small feature that can enable you to maximize the per-minute rates you may receive from your local, long-distance, and VoIP providers.

Pattern matching allows the Asterisk to identify the sequence of numbers you're dialing and treats the calls differently based on the type and quantity of digits dialed. Pattern matching is based on a specific syntax that uses the following rules:

✔ All patterns begin with an underscore (_). The underscore alerts Asterisk to the fact that you are establishing a pattern.

✔ The letter N represents numbers ranging from 2 to 9.

✔ The letter X represents numbers ranging from 0 to 9.

✔ The letter Z represents numbers ranging from 1 to 9.

✔ Individual numbers are accepted as single digits, except if they are listed with a hyphen. For example, [13-5] is understood as digits 1, 3, 4, or 5.

✔ A period (.) matches all whole numbers (beginning at 1) as well as any quantity of numbers.

The letter abbreviations for numbers are telecom industry standards and not a manifestation of Asterisk. The telecom shorthand for an area code and the first three digits of a phone number is NPA-NXX.

The most common patterns used in Asterisk to identify outbound calling are as follows:

✔ _NXXXXXX: This pattern represents a 7-digit number that is typically dialed for local calling. Although a true local call terminates only about 13 miles from where it begins, and in some areas, you can call a destination hundreds of miles away with only seven digits, this pattern is generally used to identify local calls.

✔ _1NXXNXXXXXX: This pattern represents a 10-digit number with a leading 1. This pattern matches all calls made in the North American Number Plan, which includes calls to Canada and the Caribbean.

✔ _011.: This pattern represents all true international calls. The period at the end allows any number of digits to be dialed. This is necessary because the total number of digits required for international dialing varies from as few as 10 digits to as many as 16 digits.

If you never call internationally and want an added layer of protection against malicious hackers, you can restrict your entire system to direct all 011 prefaced calls to a hangup application. You can also limit the types of calls available to each extension, or you can use this technology to maximize the per-minute rates for a variety of carriers.

Every local and long-distance carrier negotiates unique pricing deals with every other local and long-distance carrier in America. For example, MCI may have a lot of traffic to the Bell South region of Texas, so MCI can negotiate a better rate than AT&T. Local carriers may charge more to terminate calls from long-distance carriers in your hometown than they charge you through your local calling plan. You may pay a half-penny per minute for calls in a 13-mile radius of your office from your local carrier; to terminate the same calls through your long-distance carrier may cost you as much as 15 cents per minute. The bottom line is that you have different per-minute rates available to you through each carrier, and Asterisk can help you find the best of all worlds.

The dollar value of your phone bill dictates how much time and effort you want to spend building a cost-routing feature in Asterisk using pattern matching. Every company should have the discussion of how to use pattern matching to save money. Larger companies spending hundreds of thousands of dollars a month in long-distance phone service should definitely employ it to the fullest. Start with a rough approximation of your savings and then determine if the labor required is worth it. With a potential savings of 5 to 40 percent, it makes solid business sense.

When building your least-cost routing table, the per-minute rates to Alaska and Hawaii are almost always much more expensive. If your contract for long-distance only lists an interstate and an intrastate rate, check to see whether these outlying areas are covered in the interstate rate. The same applies for all countries in the Caribbean that are also reached by dialing a 3-digit area code and a 7-digit phone number, just as if you were calling to Council Bluffs, Iowa. To differentiate the calls, you have to refine your pattern even more.

To differentiate handling calls to Alaska in area code 907 from Hawaii in area code 808 and lump the rest of your long distance into one category, you must establish the following three different pattern options:

- ✔ _1NXXNXXXXXX is for your default traffic.
- ✔ _1808NXXXXXX is to route calls to Hawaii.
- ✔ _1907NXXXXXX is to route calls to Alaska.

The good news is that patterns are identified in Asterisk by the most precise fit, not by the sequence in which you enter them into your dialplan. Any outbound call to 1+808+XXX-XXXX is always processed by the second pattern. It doesn't default to the most basic _1NXXNXXXXXX plan.

Using dialstatus

The dialstatus channel variable is automatically assigned by the dial application as the call progresses from one priority to the next in the dialplan. Failed outbound calls generally fall into one of the following three categories:

✔ Busy: The number dialed is currently on an active call.

✔ Congestion: The call has failed within the network of your local or long-distance carrier. The number you have dialed may be disconnected or cannot be completed as dialed, or a larger issue with your carrier has prevented the call from completing.

✔ ChanUnavailable: The call was unable to leave your network because no outbound channels were available. For example, you only have a four-port Zaptel card and you already have four active calls on that resource.

The variable ${DIALSTATUS} allows you to treat each type of completion issue differently. Playing a message stating "The number you have dialed is busy" is great for the Busy status, but you can overcome a ChanUnavailable status by building a secondary route choice. If the ZAP/1 channel is unavailable, directing the call to SIP/Outbound may successfully complete your call.

Using 911, 411, and 611

You must build the ability to dial 911 (emergency), 411 (information), and 611 (local phone carrier customer service/trouble reporting) into your outbound dialing context.

The code to add these 3-digit codes to the outbound dialplan is as follows:

```
exten => _9[469]11,1,Dial(${TRUNK}/${EXTEN:1})
```

The pattern matching allows a leading 9 to be dialed for an outside line that is removed by ${EXTEN:1}. The [469] identifies the individual digits that are dialed, and the 11 is the last two digits. All together, it is a compact line of code that allows you to send these calls out over your TRUNK.

Chapter 6

Adding Features to Dialplans

*B*asic dialplans cover basic needs (see Chapter 5 to set up a basic dialplan). Connecting to your carriers and establishing inbound and outbound dialplans are the foundation from which you add features. But you can do much more with your Asterisk, such as voice mail, conference calling, and name directories.

This chapter covers the additional features you need to activate within your Asterisk software. In addition to walking you through the process of setting up voice mailboxes, dial-by-name directories, and conference calling, we also give you some tricks of the trade to make the process easier.

Refining Your Code with Symbols

A variety of programming symbols used throughout the world of software are applicable in Asterisk. These symbols are useful as you're programming your Asterisk system:

✔ " ": Consider everything in the quotation marks as a single entity. If the value of a COMPANYNAME is "Smithsonian Museum" Asterisk looks for the 18-character variable value of *Smithsonian* and *Museum* along with a single space separating them. If you didn't include the quotes, Asterisk would look for two separate variables, one being *Smithsonian* and a second, unrelated variable called *Museum*.

✔ ~: Use the tilde symbol when using regular expressions to represent the `like` function in a comparison. We cover regular expressions in detail later in this chapter in the "Regular expression operators" section.

✔ ?: Use the question mark to set up a section of code that makes a decision based on the true or false analysis of data such as this `GotoIf()` application:

```
(testvalue?true:false)
```

The question mark separates the test value from the possible options. If the test value is true, the first option is executed. If the test value is false, the second option is executed.

✔ :: The colon symbol holds different meaning based on the code that uses it. In general, the colon delineates code, options, or variables. In the `GotoIf()` application, it acts to separate the true and false options.

✔ |: Use the pipe symbol to separate characters or code. You can also use it to replace commas in the dialplan in some instances. For example, you can write

```
exten => 2563192010,1,Dial(SIP/2563192010,20,r)
```

with pipes like this

```
exten => 2563192010,1,Dial(SIP/2563192010|20|r)
```

!: Use the exclamation mark to negate an expression in the general programming lexicon. It functions in a dialplan syntax to provide a default path for calls that fail to generate a match. Many dialplans require a call to meet certain criteria before it can be directed to the correct extension or `dial()` application. If a phone number is dialed that doesn't include either seven or ten digits, Asterisk uses the ! symbol to automatically route the call to a busy signal or prevent employees from dialing internationally.

Conferencing with MeetMe

There always comes a time in your life when you need to chat with a whole bunch of people at once. For example, a new product launch may require all your remote salespeople to be on a call with your vendor, the support staff for your vendor, and yourself. You can't make this happen with the garden-variety three-way calling that some phone systems provide. Presenting a conference room login and password to your vendor also forwards the image that your company is stable and established. The conference room you need to set up is done in Asterisk using the `MeetMe()` application.

For general-purpose conferencing, you can add a conference room entry to the `/etc/asterisk/meetme.conf` file like the following:

```
conf=>1111,5555,9999
```

This line specifies conference room 1111, with a PIN of 5555 and an admin PIN of 9999.

You don't need to include a PIN and admin PIN if you trust everyone on the call, but we highly recommend that you set up all your conferences with PINs. Without PINs, people can use your conference room for their own use, for as long as they want. If your conference room is enabled to dial outbound, criminals can hijack your long-distance service and make international calls that you're responsible for.

The `MeetMe()` application is actually structured for three elements in the argument. In the dialplan, enter the `MeetMe()` application of an extension with the following:

```
MeetMe([confno][,[options][,pin]])
```

The following example identifies a conference room reached by dialing 2565551212 to reach the prompt for the conference bridge, with a conference room number 1111, no options, and no PIN.

```
exten => 2565551212,1,MeetMe(1111)
```

The `MeetMe()` application has many options, from a, which sets the administrator mode to x, which closes the conference bridge after the last caller leaves. Type the following commands to find a list of your `MeetMe()` application options:

```
cd/usr/sbin/asterisk - r
show application MeetMe
```

The options for the `MeetMe()` application are case-sensitive. Be sure that you list them correctly because the difference in lowercase and capitalized options could spoil your call. For example, option d dynamically adds a conference, but option D adds a conference room prompting you for a PIN.

You can access a conference room in the following ways:

✔ If you require people outside your company to enter the call, you can design individual phone numbers in your Asterisk to point to the conference room by using the following code:

```
exten => 2563192010,1,MeetMe(1111)
```

✔ You can establish a separate conference room for local Asterisk users through the [internal] context:

```
exten => 4000,1,MeetMe(1111)
```

✔ You can configure a standard conference bridge:

```
exten => 4000,1,MeetMe(1111,cMrx,5555)
```

This conference room number 1111 with the PIN of 5555 has the following four options:

✔ c: Plays a recording that states the number of callers in the conference as you join.

✔ M: Activates the music on hold when the first caller enters the conference room and continues to play it until another caller arrives.

You must have music saved in the /var/lib/asterisk/mohmp3 directory, or your music on hold sounds more like one-hand clapping.

✔ r: Records the conference. The conference is recorded as a .wav file and named meetme-conf-rec-${CONFNO}-${UNIQUEID}.wav

✔ x: Closes the conference when the last caller hangs up.

Conference rooms need a timing source to work correctly. The Linux 2.6 kernel version has an internal timing source and fulfills this requirement by itself. Older versions of Linux need either a Zaptel card to act as the timing source or the ztdummy driver and a USB port.

Your conference rooms are not bound by context and are accessible to anyone who can dial into your Asterisk. If you have external and internal groups using your conference rooms, set up different extensions that point to the same conference room with different options for each group.

The following code lists extension 4000 with all the cMrx options and a PIN of 5555. Extension 5000 enters the same MeetMe room, but the conference isn't recorded and doesn't require a PIN:

```
exten => 4000,1,MeetMe(1111,cMrx,5555)
exten => 5000,1,MeetMe(1111,cMx)
```

By giving extension 4000 to your external MeetMe users and 5000 to your internal users, you control your exposure for external users abusing the conference room. You can also establish a set of options assigned to calls based on the calling device, channel variables, or time of day instead of just the extension dialed. The possibilities are endless.

Queuing Calls

You need to find a way to handle calls while preventing your customers from receiving busy signals. If you are using Asterisk as the basis for a calling card platform or as a VoIP service platform, busy signals are lost revenue. If a call doesn't hit the Asterisk, you can't bill for it. The Queue() application allows you to receive multiple calls for the same toll-free number into a constructed environment that can hold the active calls until they can be received.

The queuing feature cannot perform magic. If you have a single analog phone line, Asterisk can't receive multiple calls for it and hold them in queue. Your local carrier identifies the line as being in use and sends a busy signal to the person calling you. You must be able to receive multiple calls from your carrier before Asterisk can place them into a queue. This requires you to have multiple analog lines from your local carrier that allow one number to hunt through the other available lines, or you must have a dedicated circuit of some sort to allow you 24 or more available channels.

Customer service departments are common candidates for a calling queue. The ability to hold new callers until the customer service representatives are available prevents the callers from simply hearing a busy signal. As much as we all hate to be stuck in a queue, it's better than having your call rejected with a busy signal. The queue gives the perception that your company is still in business and isn't without phone service.

The Queue application relies on the construction and integration of the following basic elements:

- **Incoming calls:** The calls can enter the Asterisk from Session Initiation Protocol (SIP), analog, or InterAsterisk eXchange (IAX) connections, or any other channel type that Asterisk supports.

- **Extensions set to receive the calls from the queue:** You must set up every customer service representative as an *agent* (an extension receiving calls and capable of logging either into, or out of, a queue) in the queues.conf file to accept these calls.

- **Distribution methodology:** This aspect of call queuing requires you to determine the sequence of how the calls are sent to the agents. Your options are as follows:

    - **Fewestcalls:** This option tracks the number of calls the agents have accepted and sends new calls to the agent with the least amount of completed calls. This scenario may seem like a great idea, but if an unruly customer takes an hour of time for a single customer service rep, the rep will be slammed with calls for the next few hours as he tries to catch up to the call count of his peers.

- **Leastrecent:** This option allows calls to be sent to the agent who has been idle (without a call) for the longest duration of time.

- **Random:** The Asterisk system randomly assigns calls to any of the agents logged in to the system.

- **Ringall:** This option rings all the agents who are logged in to the system and doesn't assign the call specifically to any one agent. The first agent to answer the call gets it.

- **Roundrobin:** Each agent receives a call in a set pattern with this option, just like dealing the opening hand of cards to players in a poker game. Everyone gets the same number of cards as they are passed out across the table from left to right and back to the first person on the left.

- **RRmemory:** This option allows continuity in the roundrobin, because it remembers the last person to receive a call and begins the roundrobin from there. This may not seem that special, but it enables Asterisk to keep the sequence even if it is rebooted.

✔ **Queue entertainment and announcements:** While the callers are in queue, you can have them hear music, announcements of new products, or options to direct their call to a more specific queue.

You must configure two files and use at least one specific application to make queuing work. The files are located with all the other configurations files for Asterisk in the /etc/asterisk directory. You need to configure these files:

✔ queues.conf

✔ agentlogin() application

You must also configure the agents.conf file if your members are logging in and out of a specific queue. If you are running Asterisk for a call center, this may be necessary, but if you're configuring Asterisk for a single business, you probably don't need it.

You can create a complex voice-prompting system for your callers in queue, but you can't apply a preset application to do everything. Building in options for reaching accounting, sales, customer service, or your after-hours staff can be done quite easily, but it requires a great deal of manual coding. Every recording has to be linked to the proper extension in the correct time of day and day of week in the queue.conf file.

Prioritizing calls

When is a priority not a priority? When you are referencing call queuing. In a normal dialplan application, a priority is a number that establishes the sequencing of lines of code. A *priority* in a call-queuing application refers to the level of urgency of a call, enabling Asterisk to establish a hierarchy of incoming callers — similar to a triage system.

The urgency of a priority increases as the value of the number assigned to it increases. You can create a dialplan that offers different extensions for billing questions and account changes, or to report trouble with the product. You can assign a priority to the call based on the extension a caller enters. For example, the billing questions (that you prioritized to 0) and account changes (prioritized to 5) wait in queue, while the trouble-reporting calls (prioritized to 10) are answered first.

Here's how you set a priority of 10 using the Queue application:

```
exten => 111,1,Playback(welcome)
exten => 111,2,Set(QUEUE_PRIO=10)
exten => 111,3,Queue(support)
```

Penalizing extensions

Penalties are numerical values that you assign to agents to identify your routing preference. The penalties handle incoming calls with the ringall distribution method. In spite of the fact that you may want the entire group of extensions to be able to receive calls, penalties allow you to set the preference of which agent is to receive the call first (if that person is available).

The lower the penalty number assigned to the extension, the greater the intention for the extension to receive the call (which works the opposite of priorities). By using penalties for the extensions to receive the calls, you can direct the calls to the respective departments; if they are busy, the call can still go to agents logged in for the other departments. The technical troubleshooting person may not be comfortable taking a call for a credit card payment, but at least the customers are handled quickly and efficiently.

To get penalties to work properly, you must assign the members for the calling queue a penalty using the following code:

```
member => device/extension,penalty
```

Here's how you assign penalties to the accounting, customer service, and tech service departments:

```
member=> SIP/Accounting,1
member=> SIP/CustomerService,2
member=> SIP/TechService,3
```

Cascading queues

You can also distribute the incoming calls by means other than priorities and penalties. *Cascading queues* allow a second queue to receive overflow from a primary queue. You set a timeout for your primary queue to let Asterisk know when to overflow the calls to the second queue.

You set the timeout parameter for the specific queue in the extensions. conf file. The timeout parameter is set to 60 seconds in this example:

```
Queue(sample,t, ,60)
```

Be sure to set up the queues in your dialplan in the sequence you wish for them to receive the calls:

```
exten => 133,1,Answer
exten => 133,2,Ringing
exten => 133,3,Queue(Queue1,t, ,60)
exten => 133,4,Queue(Queue2,t, ,60)
exten => 133,5,Queue(Queue3,t, ,60)
```

Getting music on hold

Music on hold is created in the configuration file /etc/asterisk/music onhold.conf. Asterisk allows you to specify a unique music-on-hold class for each device or extension that requires it. You set up each destination channel to receive the music on hold in its specific configuration file for the desired music class.

Here's how the configuration setting is generally displayed:

```
musicclass=
```

Instead of altering the configuration file, you can establish the music class for the extension within the dialplan using the set() application in the following syntax:

```
set(Channel(musicclass)=classname)
```

Music on hold isn't already preprogrammed in the required MP3 format at 8 kHz. You must place the music files in the on-hold directory for each class of music on hold you created. Create individual folders in the /var/lib/ asterisk/mohmp3 directory for each class of music, and then save the MP3 files. The choice of classes is entirely yours. In the absence of files in the directory, your caller hears silence while on hold instead of music.

Using the GotoIf() Application

Asterisk is more than capable of making decisions for you if you simply tell it the parameters of what it should do. The entry-level form of complex decision making built into Asterisk is the classic If-Then statement through the GotoIf() application.

The syntax for the GotoIf() application is as follows:

```
GotoIf(expression?destination1:destination2)
```

The application is useful if you need to make dialplan decisions based on variables. (We cover variables in Chapter 5.) You can use any variable as the basis for the expression in the GotoIf() application. Feel free to use any of the following:

- ✔ Current Call Detail Record (CDR) values
- ✔ Caller ID values
- ✔ Channel variables
- ✔ Global variables
- ✔ Variables based on values from databases other than Asterisk within your Linux environment

This application is integral to establishing a dialplan that can think for itself. For example, you can establish an ldcalling variable in the dialplan that is set as ldcalling=1 for calls entering from a specific VoIP device. You need Asterisk to handle these calls in a unique manner, so you have to test for the channel variable named ldcalling being equal to 1 with the dialplan, as follows:

```
exten => _1NXXNXXXXXX,1,GotoIf($["${ldcalling}"="1"]?2:3)
exten => _1NXXNXXXXXX,2,Goto(outbound,${EXTEN},1)
exten => _1NXXNXXXXXX,3,Playback(callnotallowed)
exten => _1NXXNXXXXXX,4,Hangup
```

The first line receives the call and establishes the routing of the call based on the `ldcalling` variable. The `GotoIf()` application works like this: If the value of the `ldcalling` channel variable equals 1, go to priority 2 of this dialplan. If the value of the `ldcalling` channel variable doesn't equal 1, go to priority 3. Priority 2 sends the call to a valid outbound extension and the call completes. Priority 3 plays a message that the call is not allowed before sending the call to priority 4, where it is disconnected.

The `ldcalling` variable is set using the `setvar=` setting. This is established in the setting for your device context. Setting calls to the `ldcalling` variable from a VoIP device called `WILEY` is done as follows:

```
[WILEY]
setvar=ldcalling=1
```

Adding Voice Mail

Voice mail is one of Asterisk's features that most people use on a daily basis. Everyone in your office needs a configured voice mailbox, as well as generic voice mailboxes for departments such as sales and customer service.

You must have an extension configured in your outbound context to receive incoming calls from people who want to go directly to their voice mailbox. You can have calls routed directly to voice mail by creating an extension terminating at the `VoicemailMain` application:

```
exten => 9999,1,VoicemailMain()
```

This code sends you directly to the voice-mail application that asks you for your extension (that is, your voice-mail extension, not your internal extension — they may be different) when you dial 9999 from your phone. This is helpful because it allows you to dial directly into the voice-mail system to retrieve your messages instead of coming in through a roundabout method.

A sample voice-mail configuration for a handful of voice mailboxes, extensions 1000 to 5000, is listed as follows:

```
[default]

1000 => 1234,George
        Washington,gwash@myvoip.com,,attach=yes|saycid=
        yes|review=yes

2000 => 5678,John
        Adams,jadams@myvoip.com,,attach=yes|review=yes

3000 => 9012,Thomas
        Jefferson,tjeff@myvoip.com,2125551212,saycid=ye
        s|review=yes

4000 => 4321,James
        Madison,jmadi@myvoip.com,,attach=yes|saycid=yes

5000 => 8765,James
        Monroe,jmonr@myvoip.com,,attach=yes|saycid=yes|
        review=yes|operator=yes
```

The five voice mailboxes we have configured all have similarities and differences. Each voice mailbox has a unique set of options. Only the voice mailbox for Thomas Jefferson has a phone number listed to receive a page when a message is left.

The origination address used in the e-mail notifications is stipulated by you in the voicemail.conf configuration file. The e-mails have generic titles on them stating *Asterisk PBX* and your voice-mail number.

We discuss how to add options to your basic voice mailbox in the following sections.

Building a voice mailbox

Voice mailboxes are easy to establish. Start by adding a voice-mail context and extension definition in the voicemail.conf file located in the /etc/asterisk/ directory.

The voicemail.conf file is all you need if you are only using Asterisk for one company. Using Asterisk for a multicompany voice-mail system requires unique extensions to be configured linking to the individual company voice-mail contexts in the dialplan.

Here's an example of the extension using `stdexten` (*standard extension macro*) to dial the VoIP device called SIP/001. If no answer is received, the call is sent to the voice-mail application at extension 0001 of the `[atlasvoip]` context:

```
exten => 2565551212,macro(stdexten,0001@atlasvoip,SIP/0001)
```

You specify additional voice-mail contexts in the same manner as the default context:

```
[vmcontext]
vmexten => password,name,email,pager,options
```

The syntax for the extension definitions is as follows:

```
extension => password,name,email,pager,[options separated by | ]
```

You can use the following syntaxes when building your voice mailbox:

- ✓ **Extension:** The voice-mail extension.
- ✓ **Password:** The password for the specific voice mailbox.
- ✓ **Name:** The name of the person whose voice mail resides at the extension.
- ✓ **Email:** The e-mail address you are establishing that can receive the recorded voice-mail message as an attachment.
- ✓ **Pager:** The pager you want to establish that can receive notification of voice mails.
- ✓ **Options:** These are the specific options for the voice mailbox. This section allows you to establish the e-mail address listed to be sent the voice mail as a `.wav` file, or to notify the pager listed that a voice mail has been received. Remember to separate each option by the pipe symbol (`|`).

Empowering voice mail with options

Normal voice mail is boring. Asterisk voice mail is, well, not boring. You can configure each voice mailbox with a variety of options. A handful of the many options available are as follows:

- ✓ `Attach`**:** This option packages the voice mail into an audio file and e-mails it to the e-mail address listed in the line of code for the voice mailbox if it is set to `=yes`. The default is `=no`, which doesn't send the voice mail as an attachment.

✔ `Delete`: This feature either deletes a voice mail after it is e-mailed with the attach option when configured to =yes or leaves the voice mail in the system if set to =no.

The Delete option can't be used as a global setting. It must be added individually for each voice mailbox to which it applies.

✔ `Review`: This allows a caller to review the message before dropping it into the voice mailbox if it is set for =yes. If the option is set to =no, the stammering, wandering message the caller hoped to delete and try over again is automatically dropped into the voice mail before the call disconnects.

✔ `Operator`: If you've set the review option to =yes, the addition of the operator option allows the caller to press 0 and be sent to an operator at any time while recording a message in the voice mailbox.

The operator extension must be configured at the 0 extension in the `extensions.conf` file for this option to function.

✔ `Saycid`: This option takes either a =yes or =no value and reads back the caller's phone number prior to playing the message. The Asterisk system reads the date and time the message was left, and then the number the call presented in the caller ID. This is a very nice feature, especially when the person leaving a message for you either forgets to leave a phone number or if she speaks so quickly that you can't understand the call-back number left on the message.

The default for all these features is =no. If you want the default for a feature, you don't need to identify it in your dialplan. If you are interested in more voice-mail options, check out www.voip-info.org/wiki/index.php?page=Asterisk+config+voicemail.conf.

Building a dial-by-name directory

The dial-by-name directory used by Asterisk is based on the `voicemail.conf` file used to establish the individual voice mailboxes.

You're probably familiar with dialing an extension to get a voice mailbox. The `Directory()` application allows you to read the last name of the person in the `extensions.conf` file, as follows:

```
exten => 3,1,Directory(default,incoming,f)
exten => 4,1,Directory(default,incoming)
```

You can search by the person's first name by adding f to the end of the `Directory()` application as we have done in the first line of code for extension 3.

The Asterisk directory application attempts to connect to the [internal] context extension in the dial-by-name directory when the application is run. If you wish to send a call to a cellphone or remote office that isn't within your Asterisk network, you have to build special code to forward the call to an outbound context and dial the remote phone.

Keep your life simple and set your voice-mail extensions to be the same as your internal extensions. If the internal extension for your sales department is 200, also use extension 200 within the voicemail.conf file for them. You don't want to have to cross-reference hundreds of internal extensions to their correct voice-mail extensions.

Recording your personal messages

After setting up the VoicemailMain extension in the [outgoing] context and building the individual voice mailboxes, you can then personalize your voice mail. Dialing into the VoicemailMain extension presents you with a prompt to input your extension and then a password. After you enter your voice mail, you can press 1 to hear your messages, or you can proceed through the prompting to change your password or record a temporary greeting, or access the standard litany of voice-mail options.

Enhancing your voice mail with the GotoIfTime() application

Voice mail is one of those features that you want to have available whenever you aren't available (including evenings, weekends, and holidays). You can also use this application to adjust how your after-hours calls are handled internally, or have an after-hours service that receives your calls when the office is closed. The application can be integrated for either case and has this syntax:

```
GotoIfTime(time, days of week, days of month, months?label)
```

Valid data for the elements of the argument are as follows:

- **Time:** Military time from 00:00 to 23:59. Asterisk doesn't deal in seconds. If you want a specific time frame, say from 9 a.m. to 5 p.m., input the data as 09:00-17:00.

- **Days of week:** The days of the week are listed in 3-digit abbreviations based on the first three letters of the day of week. If you want the application to route your calls differently on Monday, Wednesday, and Friday, input the data as Mon,Wed,Fri. If you want the application to take effect all week, list the data as Mon-Fri.

✔ **Days of month:** This option accepts digits from 1 through 31 and refers to the numerical date within the current, or specified, month. To apply the application to the date range of the 10th of the month through the 25th, write it as 10-25. To use the GotoIfTime() application for the 1st, 3rd, and 5th of the month as well as the range 10th through 25th, input the information as 1,3,5,10-25.

✔ **Months:** The months are listed with 3-digit abbreviations just like the days of the week. If you want the days of week or days of month to only be applicable for January through April, input the value jan-apr. If the application was for an event in July and September, input it as jul,sep.

✔ **Label:** A label is a name that represents a priority in a dialplan. The label can be any of the following:

- A named priority in the same extension. Each extension priority can be names that contain the label wiley:

```
exten => 2563192010,1(wiley),Answer
```

- A valid priority in the extension, such as a priority 2.

- A valid extension and priority within the context, such as 133, with 1 representing extension 133 and priority 1.

- A completely different context, extension, and priority, such as Outbound,133, with 1 representing a completely different context, extension, and priority.

The following code demonstrates how a label with a named priority functions:

```
exten => 2565551212,1(start),Dial(SIP/0001)
exten => 2565551212,2,Goto(start)
```

The start label represents priority 1. The second line of code redirects the call to the first priority in the context. This is a very counterproductive dialplan because it sends the caller into an endless loop, but it effectively demonstrates how the label functions.

Diving In with Macros

Where variables allow you to substitute one value for another, *macros* go that extra mile and allow you to substitute one phrase for an entire section of code. Macros eliminate redundant code as well as provide a means for easily adding features and updating your dialplan.

As wonderful as macros are, they are quite particular. To work properly, macros must meet these parameters:

✔ Context macros must begin with `macro-`. This macro sets up voice mailboxes:

```
[macro-stdexten]
```

✔ Marcos use the `s` extension in the Asterisk inventory. See Chapter 5 to find out more about the `Start` extension. The purpose of the macro is to take action without waiting for responses, so the immediacy of the `s` extension is natural for the macro.

✔ Macros are frequently tied to the extensions triggering them. This code refers to the value of the dialplan argument that triggers the use of the macro:

```
${ARGn}
```

The n suffix that follows `ARG` identifies the number of the priority. The priority gives you continuity with the original dialplan and the ability to integrate data from it into the macro.

This is a standard dialplan to which we added a macro:

```
[macro-stdexten];
${ARG1} - Extension
${ARG2} - Device(s) to ring

exten => s,1,Dial(${ARG2},20)
exten => s,2,Goto(s-${DIALSTATUS},1)            ; Jump
        based on status
        (NOANSWER,BUSY,CHANUNAVAIL,CONGESTION,ANSWER)

exten => s-NOANSWER,1,Voicemail(${ARG1},u)      ; If
        unavailable, send to voicemail w/ unavail
        announce
exten => s-NOANSWER,2,Goto(default,s,1)         ; If they
        press #, return to start
exten => s-BUSY,1,Voicemail(${ARG1},b)          ; If busy,
        send to voicemail w/ busy announce
exten => s-BUSY,2,Goto(default,s,1)             ; If they
        press #, return to start

exten => _s-.,1,Goto(s-NOANSWER,1)              ; Treat
        anything else as no answer
exten => a,1,VoicemailMain(${ARG1})             ; If they
        press *, send the user into VoicemailMain
```

We could also have used `${MACRO_EXTEN}` for `ARG1` instead of simply linking it to the dialed extension. The rest of the dialplan is very straightforward with the s extension. The `Goto()` application establishes the method by which the specific recording and treatment for the call are handled based on

the status of the call. The DIALSTATUS is identified by Asterisk, and the following lines of code direct the call based on that information. If the caller
receives NOANSWER, the call is forwarded to the voice mailbox for the extension dialed.

The b and u listed in the (${ARG1},b) and (${ARG1},u) arguments are
established Asterisk code for this application, representing unavailable (u) or
busy (b) parameters for the voice-mail application. Asterisk also allows the
caller to press the pound (#) button and be sent to an s extension, or the
asterisk (*) button to be sent to an a extension. This allows you to set up
code to respond to them without expressly telling Asterisk to listen for them.

After you create the stdexten macro, it appears in your dialplan as the
following:

```
exten => 2565551212,1,macro(stdexten,0001@atlasvoip,SIP/0001)
```

Make it a point to use macros often, even within a macro. The dialplan body
of a macro is constructed just like any other dialplan. If you find yourself writing the same section of code repeatedly, even within a macro, turn your
repetitive code into a macro and save time.

To use local variables in macros, you link the extension number dialed to a
macro by using the value only syntax, as follows:

```
${MACRO_EXTEN}
```

This allows you to generically link many extensions to the macro. Each extension dialed can be routed to a different application, context, or priority. To
tag a call received into a macro with the context or priority from which it
came, use the following code:

```
${MACRO_CONTEXT}
${MACRO_PRIORITY}
```

Making Expressions

Expressions are intelligent pieces of code allowing you to use simple logic.
They use variables, values, and operators (covered in detail further in this
chapter) to make decisions within your dialplan. The following syntax for the
command is very straightforward:

```
$[expression]
```

The simplest expressions to understand are the ones using simple mathematical expressions. The following code identifies some very basic expressions available in your dialplan:

```
$[3 + 5]
$[27     -     17]
$[21 / 3]
$[5   *    7]
```

Expressions can add infrastructure to your Asterisk. Counting the number of times a person misdials an extension is a good demonstration of expressions. By creating a dialplan that allows five incorrect attempts before it sends the caller packing, you reduce the amount of time wasted in your system by the caller. The code for this expression is as follows:

```
[ivrmain]

exten => s,1,answer
exten => s,2,Set(trips=1)
exten => s,3,GotoIf($[${trips}=5]?4,5)
exten => s,4,Goto(9999,1)
exten => s,5,Playback(enter-your-extension)
exten => s,6,WaitExten

exten => 1111,1,Do Something
exten => 2222,1,Do Anything
exten => 3333,1,Do Nothing
exten => 9999,1,Playback(you-waited-5-times-goodbye)
exten => 9999,2,Hangup()
exten => i,1,Set(trips=$[${trips} + 1]
exten => i,2,Goto(s,3)
```

The second-to-last line of code identifies the number of trips the caller has made to extension i as 1+ the value of trips set in priority 2 for the s extension. This allows the dialplan to add to the set number of trips until it equals five, triggering the call to be sent to extension 9999 priority 1, which tells the caller that he is being disconnected. The recording of "you-waited-5-times-goodbye" is a unique recording made for this specific application.

Using operators

Operators allow you to validate how an expression is functioning or integrate the results of an expression into your dialplan. The operators come in a variety of forms, each with their own specialty. Their pervasiveness in your

Asterisk coding depends on your personal preference and style. Some operators are easily applicable in every Asterisk application, and others are only useful to Asterisk systems that require specific functionality. Choose the ones that work the best for you.

Operators are programming expedients that compare and qualify, not only numbered values, and phone numbers, but also variables, dialstates, and names. Any piece of information that is consistently received in the same syntax can be used to group, process, or qualify.

Logical operators

Logical operators are some of the more widely applicable operators. They allow you to qualify calls based on criteria.

Using the logical AND operator

The logical AND operator evaluates the two expressions and returns the first value if both expressions in the operator are true. The standard syntax for this is as follows:

```
expr1 & expr2
```

The operator in the following code validates that a call is a long-distance call *and* that it has been listed with the channel variable of ldcalling:

```
exten => 1NXXNXXXXXX,1,Set(thisisanldcall=1)
exten => 1NXXNXXXXXX,2,GotoIF($[${thisisanldcall}&
        ${ldcalling}]?4,3)
exten => 1NXXNXXXXXX,3,Hangup
exten => 1NXXNXXXXXX,4,Dial(Zap/0/${EXTEN})
```

If both criteria are met, the call is sent to the first priority listed. In this case, it is priority 4, where the call is dialed out of the Zap card. If both criteria are not met, the second listed priority (3) is chosen and the call is hung up.

The variable ldcalling is a channel variable that can be set at any point during a call. It can be set when the call is initiated based on the channel from which it originates, or the extension it is sent to for termination. Establishing channel variables like ldcalling allows you to have Asterisk process your calls logically and efficiently.

If either of the criteria aren't available — for instance, if they aren't established with a set() application — the value either appears as 0 or populated with !expr, indicating the missing expression. In that case, the & operator chooses the second priority listed for a failed match.

The exclamation point (!) identifies the expression as holding the opposite or negative of the listed expression. In normal Asterisk coding, a response of 1 represents a positive or true match and a 0 represents a negative or false match. Following the logic that the negative of a negative is a positive, a match of !0 indicates a true response and !1 is false. Gosh, you have to love those double negatives.

Using the logical OR operator

The logical OR operator compares two expressions, presenting expression1 to be used in the dialplan if it is valid (not an empty string or 0). If the first expression is invalid, the operator then validates and uses the second expression as long as it is valid. The syntax for an OR operator is as follows:

```
[expr1 | expr2]?true:false
```

Comparison operators

Comparison operators perform a function just like their name implies. They allow comparison information to be drawn from two expressions and fall into two categories.

Matching expressions

Matching expressions attempt to establish whether two expressions are similar. If the first expression equals the second expression, the operator resolves the expressions as true and takes the appropriate action. This is the most basic comparison operator and one that you probably use frequently. The syntax is as follows:

```
expr1 = expr2
```

The operator enables you to use information you have tagged on a call with the set() application and handle it accordingly. If you have an account codes on a call, you can route them according to the code you have applied. An example is as follows:

```
$[ "${CDR(accountcode)}" = "atlasvoip" ]
```

This code shows the comparison equal (=) operator at work. The section of code identifies the account code listed in the CDR for the call and compares it with the account code atlasvoip. If the account code listed in the CDR is also atlasvoip, the expression is valid and the rest of the dialplan identifies how to process the call. It could be routed to a specific extension with a GotoIf() application or a handy-dandy macro.

Unmatching expressions

The next logical comparison expression is the test to see whether two expressions are dissimilar. We use our favorite negative coding symbol, the exclamation point (!), to create the code to identify that `Expression1` is not equal to `Expression2`:

```
expr1! = expr2
```

Determining greater or less than, as well as equal

Aside from straight comparisons, Asterisk can also resolve expressions based on whether expression 1 is greater than, less than, greater than or equal to, or less than or equal to expression 2. The code for the features is as follows:

```
expr1 > expr2
expr1 < expr2
expr1 >= expr2
expr1 <= expr2
```

These expressions also work with letters as well as numbers. Letters increase in value as you proceed from capital *A* to lowercase *z* in the Asterisk world. For example:

```
A<B, a<b, A<a
```

Mathematical operators

Mathematical operators aren't commonly used. Their main application is in resolving two numerical expressions, resulting in a new expression that is used. Mathematical operators run the gamut of mathematical applications and include the following:

- ✔ +: addition

- ✔ -: subtraction

- ✔ *: multiplication

- ✔ /: division

- ✔ %: Percentage of, whereby a percentage is derived that answers the question "Expression 1 is what percentage of expression 2?"

Always input spaces in the bracketed section of your operator code if you want an activity to take place. Adding too many spaces is okay, because Asterisk ignores them. Just don't forget to have at least one space between the elements of your equations to ensure that you get a result.

This expression is a single unit of 3+5:

```
[3+5]
```

The second expression is a mathematical equation and represents 8:

```
[3 + 5]
```

Conditional operators

The conditional operator allows you to make decisions based on the following mind-set: "If expression 1 is true, use expression 2 in the dialplan. If expression 1 is not true, use expression 3." The syntax for the code is as follows:

```
expr1? expr2 :: expr3
```

Regular expression operators

Regular expression operators integrate the use of regular expressions to resolve the operator. These expressions are generally used to compare an expression in your dialplan to a separate database you would have for just such a purpose. The syntax for using regular expressions is as follows:

```
expr1 : expr2
```

The `expression1` in the operator identifies the Asterisk element being compared to `expression2` that is the regular expression. Regular expressions aren't commonly used in Asterisk, but they are available if you need them.

A regular expression, also referred to as *regex* or *regexp,* is a standard programming expedient that finds a specific sequence of letters and numbers. Regex expressions are dynamic tools for searching data with fixed beginning and ending elements. To find out more about regex expressions, visit www.regular-expressions.info.

Realizing the sequencing

Do you remember being introduced in math class to complex equations like this one?

```
5(10-6)
   10
```

To arrive at the correct answer, you have to perform the subtraction, multiplication, and division in the correct order. If you multiplied 10 by 5 before you subtracted 6 from it, you arrive at a completely different — and wrong — outcome. Coding in Asterisk follows a similar hierarchy. The arguments are resolved in the following sequence:

1. All actions within parentheses are resolved before the results are used in the larger context of activities outside the parentheses.

2. Asterisk processes unary operators, such as !, second. Unary operators are equations such as ! or +A that only involve one valued element. This differentiates them from regular addition and subtraction operators that are processed later.

3. The colon (:), the equals (=) sign, and the tilde (~) are processed third.

4. Multiplication (*), division (/), and percentage (%) operators are processed fourth.

5. Addition (+) and subtraction (–) operators are processed fifth.

6. Comparison operators (=, !=, <, >, <=, >=) are processed sixth.

7. Logical operators (| and &) are processed seventh.

8. Conditional operators (?) are processed last.

Keep this order in mind when you are writing code. If the dialplan isn't working correctly, you may have forgotten that the ! takes precedence over multiplication.

Unless you're performing intricate mathematical calculations, you won't use most of these arguments. So many different types of operators exist that we can't list every one or give you a Web site to find everything you need. And unfortunately, operators are also applicable for UNIX, C++, Java, Visual Basic, and most programming languages. We suggest that you search Google for the system and specific operator that interest you. Searching for "Asterisk Unary" can give you a list of Web sites with solid information and avoid most of the information on unary operators for other programming languages.

Having Fun with Functions

Asterisk functions are similar to applications in that they have arguments and perform specific tasks. The way that they differ from applications is that functions perform much more granular tasks. The dial() application can send any call to a specific extension. The BLACKLIST() function identifies a specific call based on the origination caller ID to be removed from the standard dialplan for special handling.

Just like applications, the functions can be written normally or can represent the value only by containing the function in braces and preceding it with a dollar sign, as follows:

```
Function_Name(argument)
${Function_Name(argument)}
```

Over 60 functions are available in Asterisk, and the number is continually growing. They range from AGENT() to VMCOUNT(), and each has a specific task that it executes. Type the following command in the command-line interface (CLI) to find a full list of functions available on your kernel of Asterisk:

```
Show applications
```

If you want a more detailed view of a specific application, you can use the following command (replacing *XXX* with the specific application in which you are interested):

```
Show application XXX
```

Functions allow very exacting treatment to individual calls. You probably won't need them for your normal day-to-day telecom needs, but be aware of the fact they are available to you. You may never know when you might need to blacklist someone.

Chapter 7

Building Dialplan Infrastructure

Chapters 5 and 6 cover the elements necessary to perform specific day-to-day functions you require from your Asterisk. These features are wonderful to establish a functioning system, but true stability only comes with a solid supporting infrastructure.

In this chapter, we outline some essential housekeeping features we recommend building into your Asterisk — or at least knowing well enough to use them when the need arises. You may not want to monitor every call running through your system, but you may need to do this on occasion if you begin experiencing call congestion or call failures. If you are reconciling your carrier invoice with the calls you know were made on your Asterisk, the task is easier to accomplish if you can create a trail of breadcrumbs with your calls that can identify, link, or summarize them. We also show you how you can pull information from the Asterisk Database into your dialplan, or simply use Asterisk Database information for analysis, troubleshooting, and billing.

Monitoring Your Channels

Channel monitoring can mean two different things: actually recording your channels or merely viewing the information from your channels.

Channel monitoring is primarily used to troubleshoot calls. You can forward the WAV file to your carrier as substantiating data to help the resolution process along. You can also use channel monitoring to validate whether a call is hung. A *hung* call exists when all the hardware that's monitoring a call believes it to be active, but in fact the parties on the call have already hung up. A monitor with no audio on it indicates that the call was hung and isn't simply a 1,200-minute-long modem connection or a person trying to get into the Guinness Book of World Records for the longest continuous phone conversation.

Asterisk channels have complex and dynamic names. They start with the initiating device and end with a unique set of digits. Port 1 on a Zaptel card appears as Zap/1-*XXXX*, or for a VoIP device, SIP/207.111.777.39-*XXXX*. The channels on Asterisk are dynamic, potentially receiving a new series of end digits every time a new call is made from the same channel.

Using true channel monitoring

You can engage in *true channel monitoring* (recording your channels) by using the monitor() application. This application triggers within Asterisk before the call reaches the channel device (VoIP, analog, or digital), so it works the same way on all varieties of calls. The monitor must be built into your dialplan to record the call prior to being sent to the final extension by Asterisk. A sample dialplan to record a channel is as follows:

```
exten => 2565551212,1,Set(MONFNAME="${CALLERID(number)}-
        ${EXTEN}-${STRFTIME(,,yyyy-MM-dd-hh-mm-ss)}")
exten => 2565551212,2,Monitor(wav|${MONFNAME})
exten => 2565551212,3,Dial(SIP/${EXTEN})
```

This dialplan works like this:

1. It identifies the inbound call to the number 2565551212, tagging the call with the channel variable of MONFNAME.

2. This first priority logs the caller ID number, the dialed number, and the date and time the call was made.

3. The second priority starts monitoring the call established as MONFNAME, specifying that the audio is to be recorded as a WAV file.

4. The call is then connected to the number 2655551212 after the monitor has begun.

The call is recorded for as long as it is active. After the call is concluded and one of the parties hangs up, the recording stops. The recorded WAV file is saved in the `/var/spool/asterisk/monitor/` folder.

You can target a specific extension, such as 2565551212 in our dialplan, or a call queue. The specific channel that you record on is entirely up to you. This application works with any technology used by Asterisk — whether it's VoIP, analog, digital, or InterAsterisk eXchange (IAX) — because the monitor is engaged before the call is terminated to the required destination.

Showing your channel info

The second definition of channel monitoring is less "Big Brother," and we recommend including it in your standard toolbox of diagnostic code. This interpretation of channel monitoring allows you to answer the questions "What are my channels doing on my Asterisk? Are they in a busy state? Are they idle? Are they in congestion?" In this instance, *channel monitoring* refers to your ability to view the channel state of a particular call. Is the call dialing, in a busy state, or in a state of congestion?

This task isn't one that is constantly running on your Asterisk as much as it is something you use when you need it for testing or debugging. Execute the application from the command-line interface (CLI) of Asterisk with the following command:

```
show channels
```

This command lists all the channels or all the active calls handled by your Asterisk at the time. If a specific channel interests you, you can request additional information on it by using the following command (replace the *XXXXX* with the name of your specific channel):

```
show channel XXXXX
```

This command gives you a snapshot of the disposition of your circuits at the moment the command is run.

To view a progression of the states of your channels, you must execute the command every few seconds to gather an overview of how your channels are interacting and how the calls are progressing. We have executed this command for our VoIP channel named `SIP/207.111.777.39-b6e27058`, as follows:

```
voip*CLI> show channel SIP/207.111.777.39-b6e27058

  -- General --
            Name: SIP/207.111.777.39-b6e27058
            Type: SIP
        UniqueID: 1155701732.78404
       Caller ID: 2565551212
  Caller ID Name: 2565551212
     DNID Digits: 7004141
           State: Up (6)
           Rings: 0
    NativeFormat: 4
     WriteFormat: 4
      ReadFormat: 4
1st File Descriptor: 53
       Frames in: 2
      Frames out: 0
  Time to Hangup: 0
    Elapsed Time: 0h0m6s
   Direct Bridge: SIP/voipserver.atlasvoip.com-c04d
 Indirect Bridge: SIP/voipserver.atlasvoip.com-c04d
  --   PBX   --
         Context: incoming
       Extension: 2567004141
        Priority: 1
      Call Group: 0
    Pickup Group: 0
     Application: Dial
            Data: SIP/voipserver.atlasvoip.com/2567004141
     Blocking in: ast_waitfor_nandfds
       Variables:
BRIDGEPEER=SIP/voipserver.atlasvoip.com-c04d
DIALEDPEERNUMBER=voipserver.atlasvoip.com/2567004141
DIALEDPEERNAME=SIP/voipserver.atlasvoip.com-c04d

STACK-incoming-2567004141-1=Dial("SIP/207.111.777.39-
         b6e27058",
         "SIP/voipserver.atlasvoip.com/2567004141") in
         new stack

SIPCALLID=63dc44342e8d6cb36a7e3dd95d7a6274@207.111.777.39
SIPUSERAGENT=Asterisk PBX
SIPDOMAIN=voipserver.atlasvoip.com
SIPURI=sip:2565551212@207.111.777.39
```

```
   CDR Variables:
level 1: clid="2565551212" <2565551212>
level 1: src=2565551212
level 1: dst=2567004141
level 1: dcontext=incoming
level 1: channel=SIP/207.111.777.39-b6e27058
level 1: dstchannel=SIP/voipserver.atlasvoip.com-c04d
level 1: lastapp=Dial
level 1: lastdata=SIP/voipserver.atlasvoip.com/2567004141
level 1: start=2006-08-15 23:15:32
level 1: answer=2006-08-15 23:15:32
level 1: end=2006-08-15 23:15:32
level 1: duration=0
level 1: billsec=0
level 1: disposition=ANSWERED
level 1: amaflags=DOCUMENTATION
level 1: uniqueid=1155701732.78404
```

The first line of code is the show channel command for the specific channel named SIP/207.111.777.39-b6e27058. The rest of the information shows the processing of an incoming call from 2565551212 to 2567004141. The General and PBX sections of the report identify the origination phone number, the dialed phone number, and the devices attributed to each. The STACK section refers to the following dialplan specifics for the call:

- Context
- Extension
- Priority
- Application
- All additional information and arguments

The information is referred to as a *complete stack* if it contains all of these elements. The section of the data preceding the STACK section is one place where you can always find a complete stack for the call.

The last two sections of data identify the VoIP signaling and Call Detail Record (CDR) information for the call. Because Asterisk uses the Session Initiation Protocol (SIP) to send and receive VoIP calls, the section refers to the SIP specifics of the call. The CDR identifies the start time, stop time, and duration of the call. In this instance, the call stopped the same second it began and has a total duration of 0 seconds. See the next section for more info on this part of the dialplan.

Checking Your Call Detail Records

Call Detail Records (CDRs) are the historical records of every call attempted on your Asterisk. Each line of CDR holds all the vital information for a specific call. If you don't want to spend the time to use CDRs, they simply consume available memory. Why let precious memory go to waste?

If you use Asterisk for your internal company calls, you want to check your CDR for the following two reasons:

- **Validating your long-distance phone bill:** If your Asterisk shows that you make 50,000 outbound calls on your phone lines for the month and your carriers send you an invoice for 185,000 calls, have your carriers investigate your invoice.

- **Troubleshooting your calls:** An Asterisk set up with a VoIP outbound carrier, a local carrier on analog lines, and a long-distance carrier on a digital T-1 circuit can make troubleshooting a challenge. Your CDR identifies the channel and type used to terminate the call, allowing you to call the correct carrier to resolve the issue. Calling your VoIP carrier for a call that Asterisk is sending out over your T-1 line with your long-distance carrier wastes time and frustrates everyone. Check the CDR before you make the first troubleshooting call.

If you use Asterisk as the basis for telephony service provided to your customers, you can use the CDR to bill the calls to your specific customers. In this event, you need to hire an in-house programmer to add the additional software required to parse out these calls.

The CDR is located in a comma-separated file in the /var/log/asterisk/cdr-csv/ folder. All CDRs are inputted into one file. Each call is listed as an individual line item, so the file can be cumbersome if your Asterisk handles a high volume of calls.

Your CDR probably looks like a run-on sentence. It contains 17 pieces of information separated by quotation marks and commas. When you see two commas side by side with nothing between them, this means that section of data has nothing in it. The following is a line of a CDR from Asterisk for a single call:

```
"atlasvoip","2563192010","2565551212","outbound","""Brady
      Kirby"" <2563192010>","SIP/2563192010-1-
      bc96","SIP/voipserver_atlasvoip_com-9163","Hang
      up","","2006-08-18 11:09:56","2006-08-18
      11:10:11","2006-08-18
      11:10:38",42,27,"ANSWERED","DOCUMENTATION"
```

All the data piled up on itself is rather intimidating. If you import it into a program such as Textpad, it doesn't look so scary — just monotonous. We recommend opening the CDR file in Microsoft Excel if it contains less than 65,636 records; use Access if it has more records. If you are adventurous, just view and manipulate all the CDRs in Structured Query Language (SQL). In SQL, you can attach column headings to help you understand the CDR.

Breaking down the CDR to its constituent elements reveals the methodical dispensation of information. Starting from the first element to the last, the data in the CDR is as follows:

- ✔ **Account code:** Account codes are names or numbers you brand to phone calls with the `set()` application in your dialplan. This is normally done to attribute your calls to a department, individual, or project on which you're working so you can account for all phone expenses. The account code in our example is `atlasvoip`.

- ✔ **Source number:** This lists the phone number that's dialing in to Asterisk. The origination phone number dialing into our Asterisk is 2563192010.

- ✔ **Destination extension:** This identifies the destination phone number dialed by the incoming caller; in our example, 2565551212 is the destination phone number.

- ✔ **Destination context:** The destination context refers to the dialplan context containing the destination extension. Our example identifies 2565551212 as belonging to the `[outbound]` context. A call that terminates to a voicemail extension of 999 may have the `[internal]` context listed as the extension built in the dialplan.

- ✔ **Caller ID:** This is the name and phone number listed within the caller ID for this call. This section has 80 characters allocated to it. For example, `"Brady Kirby" <2563192010>` is the caller ID in our dialplan.

- ✔ **Channel:** This section of the CDR identifies the hardware device or technology originating the call in the Asterisk server. Our example lists the VoIP channel of `SIP/2563192010-1-bc96`. It could have been the first port on a Zaptel card, at which time this section of the CDR would identify `ZAP/1` as the channel.

- ✔ **Destination channel:** This is the back end of the call identifying the egress device or technology. Our data lists the server and port of `SIP/voipserver_atlasvoip_com-9163`.

- ✔ **Last application:** This is the last application engaged on this call. Our example ended in a completed call, so `Hangup()` is the last application in listed.

✔ **Last data:** This section is empty because the call was complete and no additional data was transmitted after the `Hangup()` application. If the call terminated to an outside phone number, the last application might be `dial(Zap/1)`. In that case, `dial` would be listed as the last application, and the argument of `Zap/1` for the last application would be our last data.

✔ **Start date and time of call:** This is the start date and the time the call was logged going out. In our example, the start date is 8/18/2006 and the time of the call is 11:09:56 a.m.

✔ **Answer date and time of call:** This is the date the call was answered and the time the date was logged in. In our example, the answer date is 8/18/2006 and the time of the call is 11:10:11 a.m.

✔ **End date and time of call:** This is the date and time the call ended. In our example, the end date is 8/18/2006 and the time of call is 11:10:38 a.m.

✔ **Duration in seconds:** This is the duration of the call when it existed in the Asterisk system, including the time it took to transfer the call to the extension and ring, prior to it being answered. This call had a 42-second total duration.

✔ **Billable duration:** This is the duration of the call from the moment it was answered to the moment it was hung up. Our call had a billable duration of 27 seconds.

✔ **Disposition:** This is the overall result of the call and helps to sort out your CDR and troubleshoot issues if an unnaturally high percentage of calls are failing. Our call was answered. This is reinforced by the fact that the call had a duration. Other options for the disposition of the call are NOANSWER and BUSY.

✔ **AMAflags:** DOCUMENTATION is the Automated Message Accounting (AMA) flag used for this call. We have never seen this section of the CDR used, in spite of the fact that it can be set individually per output or input device (SIP, Zaptel, IAX). As a standard rule, any call with duration is billable. These additional flags may be useful in a specific application, but we find they convolute the billing process.

✔ **UserField:** This is 255 characters of empty space. You populate this field with information identified and set in your dialplan using CDR applications such as `AppendCDRUserField()` and `SetCDRUserField()`.

Appending the CDR

You can amend simple information to the standard Asterisk CDR with the `AppendCDRUserField()` application. You may find it helpful to add the ReDirecting Number Inward Supervision (RDNIS) information to your CDR to help track calls and their path through your Asterisk. (RDNIS allows your

Asterisk to replace the original caller ID on a phone call with a new caller ID.) Add the RDNIS information to the UserField of the CDR for a call with the following code:

```
exten => 2565551212,1,AppendCDRUserField(rdnis=${CALLERID(rdnis)})
```

The `AppendCDRUserField()` application is the only application that allows you to add information to the CDR file. You can't use it to modify any other section of the CDR other than the `UserField`.

Newer versions of Asterisk offer a `cdr.conf` file for configuring aspects of your CDR. If you don't want to have CDRs generated by Asterisk, set the `enable` setting to =no. This disables the CDR logging. If you ever need to create the CDR, reset the `enable` setting to =yes.

If you suddenly realize that you need a CDR for calls generated while the `enable` setting is set to =no, you are out of luck. You can't regenerate a CDR for calls that have already been processed. Reconfiguring the setting to =yes only begins the process of creating a CDR for all new calls, but it cannot retroactively build a CDR.

The Asterisk `cdr-custom` directory is for files created from the `cdr-custom.conf` specification in the `config` file. The `cdr-custom` directory is a special-use file, not the standard CDR. If you don't like the order of the generic CDR, or you need more — or different — data, you can adjust it all and receive your custom CDR in this folder.

Integrating MySQL into your Asterisk

Database integration of the CDR file is essential for high-call-volume companies. The constant aggregation of every call passing through your Asterisk creates such a large file that it becomes cumbersome.

MySQL is the most common database system to integrate with Asterisk. Your Asterisk creates MySQL modules automatically if you have MySQL on your server when you download and compile the `asterisk-addons` file (one of the files required to compile Asterisk; see Chapter 2).

The MySQL modules created at the time of compilation give you the following features:

- ✔ **The ability to integrate the MySQL database into your Asterisk dialplan:** This allows your dialplan to use information within the MySQL database to establish channel variables for active calls. The calls are then routed based on information listed in the specific MySQL database referenced in the dialplan.

✔ **A real-time module that is created within MySQL:** This allows interaction with Asterisk in real-time architecture.

✔ **A CDR add-on and configuration file:** The add-on configuration file for MySQL is located in the `configs` folder with all the other configuration files. First, copy the file to the `/etc/asterisk` configuration files directory, and then modify it to include the following information for your MySQL database:

- Hostname
- Username
- Password
- Database
- Port

After you add all the pertinent information to your configuration file, you need to manually add the database file to hold your CDRs. If you are unsure of how to create the necessary database, a wealth of information is available at `www.mysql.com`.

Even an integrated Asterisk CDR file working with MySQL continues to add CDRs to the same file. For example, the file in our Asterisk holds over a year's worth of CDRs, consisting of 711,257 individual call records. That number of calls consumes 164MB. Even if you had ten times as many calls, your total file size usage is still less than 2GB.

Working with the Asterisk Database

The Asterisk Database (AstDB) isn't your normal relational database like Access or SQL; it is a slightly different creature. It's an internal assemblage of keys and values that encompass everything in your dialplan's form (contexts to extensions, devices, and variables). It holds and uses all these elements, but you can't export a preset query or view a generic report, because none exists. You don't need to know the molecular structure of it, just the ins, outs, and common uses.

AstDB uses the following two elements to organize its data:

✔ **Families:** These represent the overall group in which the data is associated.

✔ **Keys:** These are the lower-level delineations for data within the families.

The following code shows how to assign a family in AstDB named `sample` with the keys of `count` and `privacy`:

```
DB (family/key)
```

To reference the `sample` family, use the following code:

```
DB (sample/count)
DB (sample/privacy)
```

The keys generally reference a database for an application. The key could reference a single application, such as `Count()` or a group of similar applications represented by `app_XXXX.c`. For example, `app_meetme.c` would indicate all the `MeetMe()` applications: `Meetme()`, `MeetMeAdmin()`, and `MeetMeCount()`.

You can use a key only once in each family. Any information set to an existing key in a family with the `set()` application overwrites the previous data. Use the following command to determine whether a key is already in use:

```
DB_EXISTS(family/key)
```

Asterisk responds with 1 if the key exists in the specified family and 0 if it doesn't exist. It is a simple test, but it's very useful to prevent overwriting sensitive or hard-won data.

AstDB is a local storehouse for data used by the Asterisk dialplans created on the server. If you're running a configuration with multiple Asterisk servers, you must create a means to replicate all the information within each AstDB to all the servers. An off-the-shelf method for propagating the information to all the servers doesn't exist, so you have to write some custom software to make it happen.

AstDB is most commonly used to establish internal blacklists. If certain individuals are calling your business and you don't want to speak to them, you can add them to a blacklist. A blacklist isn't a destination as much as a way of partitioning unwanted calls. Your dialplan could play a message you record stating "Don't call here. We don't like you" and hang up the call. On the other hand, you could send these calls to a 25 minute recording of the hamster dance. The choice is yours.

Setting info into the AstDB

Before you can use the information in the database, you must input it. Because the AstDB isn't a standard database that allows you to easily call up tables to input, the process is a bit more complex.

You can save information into the AstDB in two ways: a direct save from the CLI or by using something like the set() application to take information from the dialplan and assign a key value to it.

To input information directly from the CLI, use the following syntax:

```
Database put family key value
```

Using this command to input a `private` key in the `sample` family appears as follows:

```
Database put sample private yes
```

Executing the same information in a dynamic fashion within the dialplan uses the DB abbreviation to denote the AstDB. Setting the AstDB to =yes for calls to extension 133 uses the following code:

```
exten => 133,1,Set(${DB(sample/private)=yes})
```

Inputting data into AstDB

You don't input data into the AstDB by using the command line or by editing a table. Information is added to the AstDB through tools built into the dialplan.

AstDB is most commonly used to store and maintain a company blacklist. You have to set up a series of macros designed to prompt you to input the phone number that you want to add or remove from the blacklist. After the number is confirmed, the appropriate macro adds or deletes the number from the blacklist in the AstDB.

For the blacklist to work, you must also build macros to process calls found on the blacklist based on the caller ID, as well as have a method to handle calls without caller ID. This is just one example of the complexity required to add and delete information from the AstDB.

Building blacklists is beyond the scope of this book. If you want to build and maintain blacklists, check out the following Web site:

```
www.jackenhack.com/blog/archives/2005/09/26/adding-blacklist-to-an-asterisk
            home-pbx-voip-server/
```

The AstDB was designed for internal use on the individual server running the Asterisk software. One of the most limiting factors with blacklists is that the blacklisted phone numbers are established as the keys in the AstDB under the family name of `blacklist`. Because you can't have a duplicate value for a family, each Asterisk server can have only one blacklist. This limitation prevents you from creating unique blacklists for each department of your company. This may not be a problem if you are using Asterisk solely for your own use, but if you're a value-added phone company using Asterisk to serve many customers, this may be a problem. One company's blacklisted irritating customer may be the mother of an executive at another company that you are also serving.

The AstDBs are specific to the server on which they are running. Creating a solid blacklist file is wonderful, but only one server knows it. If all your servers handle calls, you must forward the blacklist information to the other servers. If a call that should be blocked hits a server without the latest AstDB file, the blacklist file won't have the information and the call won't be treated as being blacklisted.

Cleaning up the AstDB

Over time, you'll probably need to delete elements from the AstDB. A person who was formerly on your blacklist might have redeemed himself; the key or family may have outlived its usefulness.

Deleting a key in a specific family within AstDB is done with the DBdel() application using the following code:

```
exten => 133,1, DBdel(sample/private)
```

To delete the entire family, use the following DBdeltree() application:

```
exten => 133,1,Dbdeltree(sample)
```

Going beyond AstDB

You probably don't need more than the standard AstDB if you are using Asterisk solely for internal company use. Asterisk can integrate with the following databases:

- MySQL
- PostgreSQL
- Microsoft SQL Server
- Oracle

Aside from database-specific pieces of integration, like the SQL syntax, one database is not vastly superior to another. Choose the database that you are most comfortable and knowledgeable with for integration. Steer clear of databases with a maximum capacity of 2,000,000MB per table, and you won't have problems with your Asterisk. You can always purge the records every few months to fend off capacity issues, but it is easier to simply use a database without such limitations. Hire a competent technician if you don't feel comfortable managing your database. Each database has its own nuances, and errors in setting up the database can result in challenges down the line, or worse, a configuration that makes it impossible to access your information.

An application programming interface (API) is an essential software component when integrating Asterisk with a larger external database program. The API could consist of nothing more than a file format set up so that your MySQL or Oracle database understands the syntax of the CDR file from Asterisk, or it can act as a programming interface for dialplan enhancements. If you're solely using Asterisk for your company, you don't need an API. If you are building a telecom empire with Asterisk as the backbone of your network, start looking for a good programmer now. The task of creating an API can be very involved. The price for consulting and programming your specific API can vary from $50 to $250 per hour.

Chapter 8

Operating the AsteriskNOW GUI

This chapter covers all the information necessary to program your Asterisk software to meet your telecom needs. We cover the basic extensions, features, and functions you'll require to serve your business. The graphical user interface (GUI) makes this much easier than using the command-line coding that the normal Asterisk requires. This allows you to build out the Asterisk software much quicker and with less potential for error.

The standard Asterisk requires typing individual lines of text into pre-existing files such as `extensions.conf` or `queues.conf`. The GUI removes the need to build all this code manually and makes the entire process much more intuitive. It establishes a progression for your installation that slowly builds out the software as you move down the menu bar.

You must have installed the AsteriskNOW software on your Linux server to use the graphical user interface. If you still need to install AsteriskNOW, turn to Chapter 3.

Connecting to the GUI

Programming your Asterisk isn't done from the Linux server on which you loaded the AsteriskNOW software; it's done from a Web portal. The specific IP address to which you connect is identified in the AsteriskNOW Console Menu (see Chapter 3). Input the full Web address, from the `http://` to the end of the IP address, into your Web browser.

A note about Web browsers and connections

You must access the site with a dedicated Internet connection. Dialup is too slow to proceed through all the security negotiations on the site, and your browser times out. Ensure that you have at least DSL or a cable modem connection before you begin programming your Asterisk.

You have many Internet browsers from which to choose. We've found the Mozilla Firefox Web browser to be the best for the job. Internet Explorer doesn't sync well with the site, and we had to log in and out several times to select additional options for programming. If you don't have Firefox on your computer, you can download it at `www.mozilla.com`.

After you enter your info, you receive a security pop-up identifying that the site is secure. Accept the security certificate, and you gain access to the AsteriskNOW site.

If you're skittish about leaving the IP address validated, your Web browser may allow you to accept the security certificate as a temporary one for the duration of your connection.

After clearing the security issues, you arrive at the login page for the GUI, as shown in Figure 8-1.

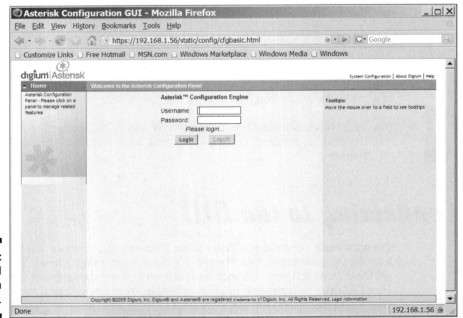

Figure 8-1:
The GUI login screen.

Enter **ADMIN** in the Username field and the password you assigned to the ADMIN username during installation in the Password field. Note that this site isn't looking for the root password for your Linux installation or your login and password for the Wikis on the Asterisk Web site.

Programming AsteriskNOW from the GUI

The main programming window for the GUI has three columns of information. The leftmost column consists of tabs. The tabs identify the elements for which you can program, starting with the Home tab and proceeding to the Options tab. They are listed in the order you'll most likely need them, with each tab down the list using programming that was accomplished in a previous tab. For instance, you configure users on the first tab and assign voice-mail options on the Voicemail tab. You then move on to putting together call queues on the Call Queues tab.

Ignoring the programming of any element as you progress down the list may result in the inability to add features and functions in later selections. So work through each tab from top to bottom to ensure that you have a completely programmed system.

Click any of the tabs on the left of the screen for a brief explanation of what's available in that section of the GUI, as well as to populate the programming options in the center section. Figure 8-2 shows the basic programming interface with the Options tab selected. The Active Settings description is listed below the Options heading, and the center section is a populated window that allows you to change your password.

The rightmost column on the GUI shows your Tooltips. Think of this as a cheat sheet for the GUI that provides explanations. If you want to know a brief description of any tab on the left, hover your cursor over the heading (you don't have to click) and you get a brief description of the heading in the Tooltips. Tooltips become helpful when programming users or queues, because it also works for elements in the center column of the GUI.

Establishing user extensions

The most basic element of any phone system is the user extensions — essentially the internal extension you must dial to reach a specific person or department. AsteriskNOW makes it easy to add extensions and builds the framework for integrating the extension into voice mail, dial-by-name directories, and call queues. Figure 8-3 shows an existing extension for Beatrice Wood (6000), a default extension of 8500 for accessing voice mail, and a new extension being created for William Spratling (6001).

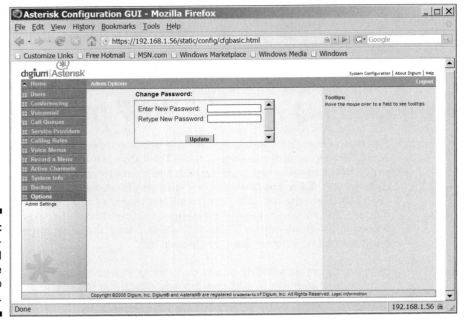

Figure 8-2:
Asterisk-
NOW GUI
with the
Options tab
selected.

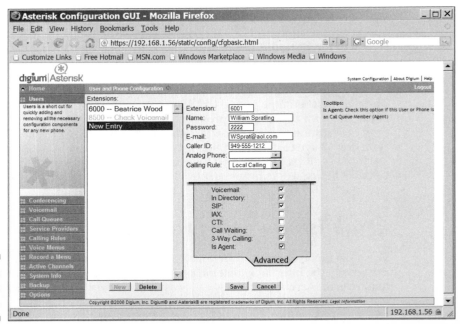

Figure 8-3:
The Users
tab.

To add a new extension, select the NewEntry option in the Extensions window and then click the New button. Fill in the following options:

- ✔ **Extension:** Assign a 4-digit extension. The system defaults to 4-digit extensions, beginning with number 6000.

  Attempting to use extensions of varying lengths at the same time could produce unexpected results. A 2-digit extension of 61 and a 4-digit extension of 6178 will cause problems. Asterisk sends the call to the 2-digit extension 61 before it receives the last two digits dialed. Keep it simple and maintain a consistent digit count for your extensions.

- ✔ **Name:** This is the first and last name of the individual assigned to this extension. It can also be a department, such as Sales or Support, for example.

- ✔ **Password:** This is the password that the owner of the extension uses to access voice mail.

- ✔ **E-mail:** Specify an e-mail address that Asterisk can send voice mails to.

- ✔ **Caller ID:** This identifies the caller ID presented when the listed extension dials out. In our case, William Spratling's outgoing calls are branded with the caller ID of 949-555-1212.

- ✔ **Analog Phone:** Identify extensions connected by analog phones if you have analog cards installed on your Linux server.

- ✔ **Calling Rule:** This references the Calling Rules tab. Based on the calling rules you've created, you can restrict the outbound dialing of this extension to local calls, emergency calls, or standard long-distance calls. You might also block or allow international (011-prefix) calls.

  Some international destinations can be reached without dialing 011. Using a calling rule that restricts 011 dialing prevents the extension from reaching Africa, Europe, Asia, Oceania, and South and Central America. However, Canada, the U.S. Virgin Islands, Guam, Saipan, and Puerto Rico, as well as a handful of Caribbean countries, are all part of the North American dialing plan and can be reached by dialing 1 plus the area code and a 7-digit phone number. If you wish to block these destinations, just block all long-distance calls from the appropriate extensions.

Click the Advanced tab to fill in more options (refer to Figure 8-3). It establishes the connections from the listed extension to other systems within the Asterisk server. They include the following:

- ✔ **Voicemail:** This option builds a voice mailbox for the extension.

- ✔ **In Directory:** AsteriskNOW establishes a directory of all extensions so that inbound callers can reach someone in your office by dialing the first few digits of the person's first or last name. Select this option if you want the company directory to include the name that is associated with an extension.

- **SIP:** Select this option if you want the extension to send and receive calls using the VoIP Session Initiation Protocol (SIP).

- **IAX:** Select this option if you want the extension to send and receive calls using the InterAsterisk eXchange (IAX) protocol.

- **Call Waiting:** Select this option if you want the extension to accept only one call before it is identified as busy if call waiting is not enabled.

- **3-Way Calling:** Select this option to allow the extension to receive a call and then dial out to another phone number to conference the inbound call, that extension, and the recipient of the outbound call.

- **Is Agent:** Call queuing is made up of a bank of agents that receive calls. An extension listed as an agent can be added to queues from the Call Queues tab (see the section "Using call queues," later in this chapter).

After filling out all the basic options and selecting the appropriate check boxes for the Advanced options, click the Save button and that extension is added to the list of extensions.

Building conference rooms

Every company reaches a point where it needs more people on a phone call than it can effectively daisy chain with three-way calling. Asterisk conference bridges allow you to do just that and also project an image of a professional company, with music on hold and the ability to record your conferences. Figure 8-4 shows the Conferencing tab.

Select the NewEntry option in the Extensions window and click the New button to design a conference bridge. Fill in the following options:

- **Extension:** The GUI auto-populates the extension with the next available extension in sequence, but you can always change it to any extension you want.

- **PIN Code:** You can assign the PIN code used by participants to enter the conference.

- **Administrator PIN Code:** Assign an administrator PIN code used by the moderator of the conference to open the conference bridge.

- **Play Hold Music for First Caller:** Selecting this option plays music for the first caller that enters a conference until another caller joins. Some people don't like to sit in a quiet room alone, and this feature prevents that from happening, even in a virtual room.

- **Enable Caller Menu:** This feature allows callers to access the Conference Bridge Menu by pressing the asterisk key (*) on the phone.

- **Announce Callers:** Select this option if you want all new callers to identify themselves as they arrive to the conference.

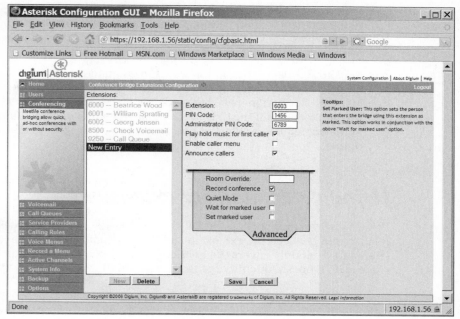

The advanced features provide some great functionality and heightened control to your conferences. If you are bringing together competing sales teams
or vendors in a conference, you probably want to keep them from chatting
among themselves before the host arrives. The following features allow you
to do that:

- **Room Override:** Input the extension number of another conference
  bridge based on another extension to enter this conference. This is helpful if you want two groups of people on a conference, but need to provide them different accessibility and features.

- **Record Conference:** Select this option to record all the conferences executed with this extension. If your business requires you to validate customer's orders by recording their approval, this option fulfills your
  requirements.

- **Quiet Mode:** You may choose this feature for a conference bridge with
  the Room Override option because it gives all users on the bridge listen-
  only access to the conference. Establishing two access points, with only
  a single group of people using the main extension and the other group in
  quiet mode, allows a controlled environment to deliver information
  while the second group listens.

✔ **Wait for Marked User:** This is a "Big Brother" feature that keeps all participants in quiet mode until a special participant, using a unique extension, arrives. Only after the marked user arrives is the audio activated so that all the participants can speak to each other.

✔ **Set Marked User:** This option works in conjunction with the Room Override and the Wait for Marked User features. Selecting the Set Marked User option makes the individual arriving from this extension the marked user. If the CEO of the company didn't want anyone chatting in the conference bridge until he arrives, you can achieve that with this option.

Using voice mail for extensions

Voice mail is an available option for every extension on AsteriskNOW. The relationship between the extension and the voice mail is established on the Users tab. That only covers the relationship between the extension and the voice mail; it doesn't identify the parameters of the voice-mail service itself.

Figure 8-5 identifies the general and advanced options for voice-mail configuration with AsteriskNOW. The following standard features satisfy the voice-mail needs for most companies:

✔ **Extension for Checking Messages:** Enter the extension you want to use to check messages.

✔ **Attach Recordings to E-Mail:** The voice-mail message can be attached as a WAV sound file to an e-mail notification.

✔ **Say Message Caller-ID:** This option prevents those frustrating moments when someone leaves you a very important message, but fails to leave his or her callback number. The Say Message Caller-ID option reads the caller ID before the voice-mail message is played.

✔ **Say Message Duration:** If someone is rambling on with a message, this option identifies exactly how long a person was speaking.

✔ **Send Messages by E-Mail Only:** Select this option to have e-mail be your only means of voice-mail notification.

✔ **Maximum Messages per Folder:** Set the maximum number of messages per voice mailbox.

✔ **Maximum Message Time:** Set the maximum duration of a message that a caller can leave.

✔ **Minimum Message Time:** Set the minimum duration of a message that can still be regarded as a message.

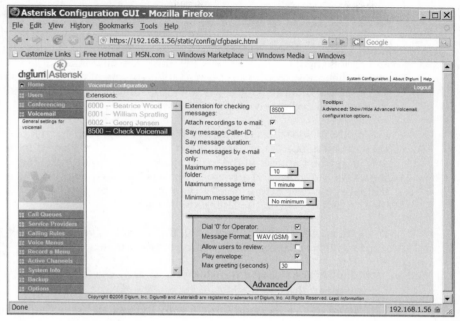

Figure 8-5:
Voice-mail
configu-
ration.

Click the Advanced tab to set the following advanced options:

- **Dial '0' for Operator:** Callers who are sent to voice mail can press 0 for the operator and be transferred while either listening to the voice-mail salutation or while recording the message. Without this option, pressing 0 is ignored.

- **Message Format:** Three audio formats are presented here; choose the default for recording all voice mails.

- **Allow Users to Review:** This option allows the incoming callers to review a message before it is saved.

- **Play Envelope:** The envelope is the date, time, and caller ID related to a voice mail. Selecting the Play Envelope option actually does the opposite of what you would imagine: It causes the envelope of the voice-mail messages to *not* be played.

- **Max Greeting (Seconds):** You can stipulate the maximum amount of time available for your employee greetings.

Using call queues

A call queue functions the same way as a normal queue: It lines up people and allows them to wait to speak to your sales, customer service, or other group of employees who are taking a high volume of calls. The feature allows you to speak to more people when it's convenient for them, instead of going to voice mail and receiving a callback when time permits.

AsteriskNOW identifies extensions under the Users tab as capable of belonging to a call queue. If the user doesn't have a check mark next to the Is Agent option, it won't appear in the list of agents on the Call Queue tab. Figure 8-6 shows the standard and advanced options for call queuing. See the section "Establishing user extensions," earlier in this chapter, to find out how to set the Is Agent option on the User Extensions tab.

The Extensions window identifies all the users you've created. Follow these steps to fill in the standard optioning for the queue:

1. **Identify the 4-digit extension you wish to use to receive calls into the queue in the Queue field.**

2. **Name the queue something catchy in the Full Name field.**

   The queue is referenced by this name, so make it descriptive.

3. **Choose a strategy for how the calls will be transmitted throughout the queue from the Strategy drop-down list.**

   Your choices are as follows:

   - **Ring All:** This option rings every agent who isn't on an active call when a new call arrives. The first agent to answer the call receives it.

   - **Round Robin:** Every available agent receives a call in turn, like cards are dealt in a poker game.

   - **Least Recent:** The agent who's been without a call the longest receives the next call.

   - **Fewest Calls:** The agent who's handled the fewest calls receives the next incoming call.

   - **Random:** This option is the luck of the draw, and any agent can receive the next incoming call.

   - **RRmemory:** This option is like round robin, but only smarter. It remembers over the course of days, weeks, or years where it left off the previous day, week, or year.

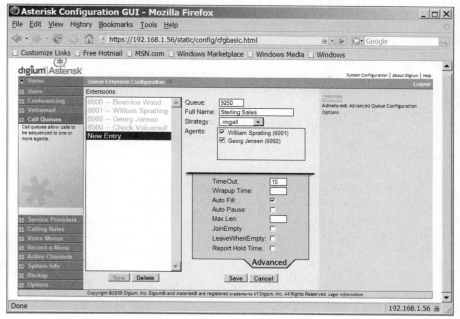

4. Assign the agents in the Agents box.

The Agents box lists all users that are designated as an agent. Many users may be listed as potential agents, but some may be assigned to a sales queue and some for a service queue. This box lists all agents, and you can choose which users you assign to the queue. In our case, we've selected both William Spratling and Georg Jensen.

The advanced options on call queuing concern themselves with the timing and managing of the calls as well as the agents. You may not want to work with these finer points of call queuing until you have an idea of call volume and the turnover of calls by each agent. The available options are as follows:

- ✔ **TimeOut:** The default on this option is 15, representing 15 seconds that an agent's phone rings before the call is forwarded to another agent.

- ✔ **Wrapup Time:** This is a buffer of time that allows your agents to finish work on one call and remain unavailable in the queue. The default on this option is 0 seconds, providing no buffer time for your agents before the next call can ring through.

- ✔ **AutoFill:** This option is enabled by default. It allows multiple calls that arrive at the same time to be immediately forwarded to agents. Asterisk does not determine which call came in first, nor does it determine the sequence of delivery.

✔ **Auto Pause:** If an agent fails to answer a call, this option temporarily postpones sending calls to the agent. It seems like a gracious option if someone is stepping out for a drink of water for a bout of the hiccups, but the default is to have it disabled.

✔ **Max Len:** This option sets the maximum number of callers allowed in the queue before the callers are sent to voice mail or receive a busy signal. The default is 0, which allows an unlimited number of calls in queue before they are sent elsewhere.

✔ **JoinEmpty:** This option allows callers to enter a queue, even if no agents are logged in to it. If this option is not selected, callers can't enter a queue until at least one agent is present.

✔ **LeaveWhenEmpty:** This option mirrors the JoinEmpty option, but it represents a queue that had agents logged in, but now have left. You can program the queue to shut down when the last agent logs out. The existing callers in the queue are forced to exit, and no new callers are granted access to the queue.

✔ **Report Hold Time:** This option tells the agent how long the call was holding in the queue before it was sent to the agent. If the hold time is long, the agent can expect a frustrated customer, but at least he isn't blindsided.

Creating trunks

AsteriskNOW assigns trunks on the Service Providers tab in three categories, two of which are VoIP.

Select the NewEntry option in the Trunks window, and then click the New button to add a trunk. The interface cards found by AsteriskNOW during installation dictates the specifics of the trunks that you can configure. If analog cards weren't found at that time, no options are listed. Figure 8-7 shows the Service Providers tab with two existing trunks and one in the process of being established.

Trunks `Trunk 2` and `VoIP` are already established and listed in the Trunks window. Both of them were established as custom VoIP trunks. The newest trunk, `LDonly`, is also a custom VoIP connection using SIP, with the Host, Username, and Password required for authentication. Click the Save button to establish the trunk and add it to the Trunks window.

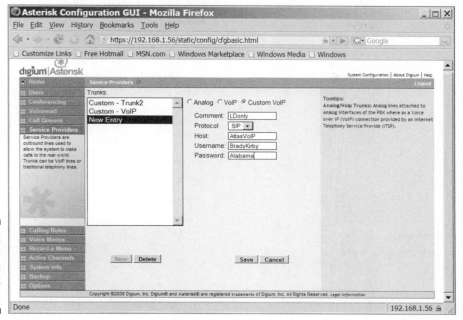

Changing the rules

The Calling Rules tab allows you to use basic pattern matching to differentiate calls and route them accordingly. Your VoIP connection may not provide service for 911 (emergency) or 411 (directory information) calls. The drop-down options allow you to choose from a selection of patterns that identify local, long-distance, and international dialing patterns, in addition to 911 and 411, and a trunk group to which the calls should be sent. Figure 8-8 shows a local calling pattern that identifies calls sent to our `Custom -Trunk 2`.

Effectively using pattern matching can save you money and minimize your frustration. Take the time to work through all the patterns listed and assign them to the correct trunks when establishing your AsteriskNOW.

Designing voice menus

While calls are processing, Asterisk runs background applications that can be used to provide additional features. The background applications monitor calls and respond to digits dialed while the call is active with another application. The AsteriskNOW GUI allows you to design these background applications. Figure 8-9 shows the Voice Menus tab.

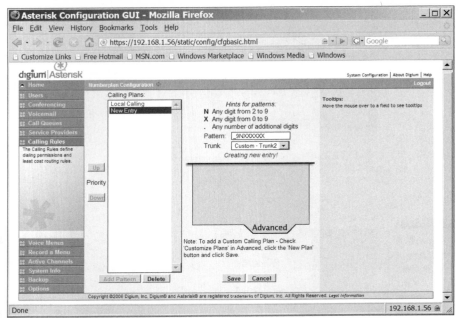

Figure 8-8:
Calling rules
with pattern
matching.

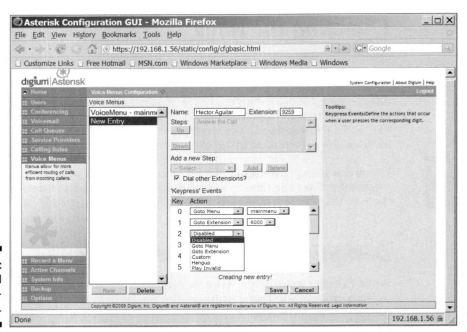

Figure 8-9:
Voice-mail
configu-
ration.

The tab begins with a Voice Menus window. Select the NewEntry option, and then click the New button to set up a new configuration for a specific extension. We've selected extension 9259 for Hector Aguilar and then specified the results of the caller dialing 0 (having the caller sent to the main menu) and 1 (sending the caller to extension 6000). The *Keypress Event* (the result of pressing a number on your telephone's keypad while on an active call with Asterisk) for pressing 2 has not yet been established, but the drop-down menu lists the options that allow you to choose how to treat the call.

Recording the Keypress Event options for your callers helps to streamline your communications. Many companies allow you a host of options when reaching an extension that can all be integrated into Keypress Events, such as the following events:

- ✔ Press "0" to reach an operator.
- ✔ Press "1" to go directly to voicemail.
- ✔ Para Espanol, markee el numero dos.
- ✔ Press "3" to be sent to customer service.
- ✔ Press "4" for sales.
- ✔ Press "5" if you're calling after hours and have a service affecting emergency.

The Goto Menu option from the drop-down list allows you to build a great deal of intelligence into your Asterisk. Building extensions for specific tasks, such as a customer service or sales queue, allows you to funnel calls to it from any other extension through the use of these Keypress Events.

The Dial Other Extensions option in the center of the window is important. This option allows an inbound caller to break out of the listed Keypress Events listed and dial another extension. On the surface, this seems harmless enough, but a technically savvy person might be able to hack through your Asterisk to find an outside dial tone and use it fraudulently. Any extensions that are known to the public should be completely handled by the Keypress Events. To protect your Asterisk from being compromised, these extensions should not allow callers to dial other extensions.

AsteriskNOW management options

The last four tabs help you manage your AsteriskNOW server and the GUI. You may or may not use them, depending on how much interaction you have with your AsteriskNOW. The last batch of tabs are as follows:

✔ **Active Channels:** This tab provides you with a remote view of the active calls and devices on your AsteriskNOW. It displays a snapshot of the activity of the server and can be refreshed to view the progression of calls.

✔ **System Info:** This tab displays the general system information of the Linux server running AsteriskNOW.

✔ **Backup:** This housekeeping tab allows you to back up your AsteriskNOW to preserve your changes.

✔ **Options:** You can change the password to the AsteriskNOW Web GUI from this tab.

Chapter 9

Utilizing VoIP Codecs

Asterisk and VoIP are a great team. VoIP provides a fast and efficient means to send and receive calls, and Asterisk coordinates them for your own internal use. The amount of bandwidth used, as well as some of the features available to you, varies based on the codec you use.

Codecs (short for *coder-decoders*) convert calls to and from VoIP. Every codec has an international standard to which it conforms for every aspect of functionality. The good news is that Asterisk is an accepted software platform for all VoIP codecs, and Asterisk doesn't need to make any special adjustments to complete calls to or from any major carrier.

Codecs are used for voice calls, faxes, and video. Each application has a variety of codecs available, each with its own costs and benefits. Some costs are simply *propagation delays* — the delay in transmission of the call due to the process of converting the signal from analog or fax to VoIP. Other costs include money that you spend for licensing your Asterisk to use the codecs.

This chapter covers the specifics of how codecs apply to Asterisk, the pros and cons of codecs, and the costs and benefits of codecs. We also cover the ability of Asterisk to handle fax and video transmissions. If you need more info about VoIP, Appendix B covers the basics.

Choosing a Voice Codec

A handful of VoIP codecs are available to choose from, but in reality, two codecs are the most popular. The uncompressed codec of G.711μ law (for North America — the rest of the world uses G.711a law) and the compressed G.729 are the codecs most widely used by VoIP carriers. Other codec options are available, such as G.726 and G.723.1, but they are generally ignored and aren't provisioned on VoIP connections.

Most carriers provision G.729 as their first codec of choice, with G.711 as the fallback. This failsafe system of using G.711 as a back-up allows the carrier to initiate the connection on the codec that uses the least amount of bandwidth (G.729) before defaulting to the uncompressed codec (G.711).

Using uncompressed G.711

G.711μ is the safest and simplest codec to use. It takes very little time to convert a call to VoIP with G.711, yielding a very low propagation delay. Some of the other, less-utilized codecs have larger propagation delays that result from the dual tasks of converting the call to VoIP and compressing the data.

Delays in the transmission of a VoIP call accumulate from where the call originates to where it finally terminates. If you build over 200 milliseconds of delay from one end of the call to the other, you expose yourself to call-quality and -completion issues caused by latency. Every bit of delay adds up, so avoid codecs with excessive propagation delay.

In general, uncompressed codecs are easier to implement because they don't require special licenses, and they are good for both voice and fax transmissions. Every carrier uses G.711 as a fallback codec to ensure that the connection completes.

The downside of G.711 is the fact that it consumes a lot of bandwidth. Due to the overhead required on the call, a G.711 VoIP call uses more bandwidth than an analog (or digital) non-VoIP call.

A standard T-1 line has 1.544 Mbps of bandwidth available. A normal, non-VoIP dedicated digital circuit is broken into 24 channels that can handle one call each. VoIP circuits aren't broken into channels; instead, they are categorized by the number of consecutive calls possible. Consecutive calls represent the quantity of active calls possible at the same time. An uncompressed VoIP circuit running the same 1.544 Mbps of bandwidth with G.711 calls can only handle about 18 consecutive calls at a time.

Compressing voice with G.729

Consecutive calls are even more important when dealing with G.729 than with G.711. For the same T-1 line of bandwidth from which non-VoIP calls can squeeze 24 calls, G.729 can easily push more than twice as many calls through the same bandwidth. The downside to the G.729 codec is the fact that you must pay a license fee to use it on your Asterisk server.

You can pay a one-time licensing fee of $10 per consecutive call directly to Digium (www.digium.com). Depending on your call volume, you could require just two licenses. The license allows you to encode/decode G.729 on a single active channel.

If you have enough Internet bandwidth and don't want to buy an inventory of G.729 licenses, simply program G.711 as your first codec of choice. This negotiates all possible calls at the uncompressed codec, so you don't need many G.729 licenses. The licenses aren't that expensive, so you should have a few for applications or customers that aren't as generous with their bandwidth.

Determining the number of licenses you need

You purchase Asterisk licenses for G.729 once and they never expire. The licenses allow Asterisk to compress and decompress the voice portion of a single G.724 call. Ten licenses installed on your hardware can handle hundreds or thousands of calls per day, as long as you don't have more than ten active calls at any time.

Asterisk doesn't proactively know whether you run out of G.729 licenses. Attempting to complete a G.729 call when all the licenses are being used results in the call failing to a busy signal. In spite of the fact that you may have more than enough bandwidth available to send the call, the call is denied, just as if you ran out of bandwidth. Because Asterisk doesn't have a gauge to tell when you are low on licenses, analyze your usage and plan accordingly. Answer the following questions to develop an estimate for the number of licenses you require:

> ✔ **How many active calls do you expect?** Tallying the potential calls handled by your Asterisk is easy if it's only used as a company-level phone system. Using Asterisk as the basis for a telecom carrier network requires more analysis. You must calculate the quantity of customers using G.729 and their current and projected call volume. Growth in the VoIP arena is so great that you may need to revisit your analysis every quarter or every month to ensure that you aren't about to be overrun.

✔ **What type of calls do you expect?** Take these types of calls into consideration:

- **VoIP–to–non-VoIP calls:** Using your Asterisk as a gateway to convert VoIP calls to analog or digital calls is one of the strengths of the system. Because only one side of the call is using the G.729 codec, you only need one license to facilitate this call.

- **VoIP-to-VoIP calls:** Having VoIP devices on both sides of your Asterisk means that you have an incoming and an outgoing VoIP connection. Each connection requires a license, so budget for two licenses on this type of call.

- **VoIP-to-application calls:** Calls that receive in VoIP but terminate in voice mail, a conference room, or other Asterisk application only have one VoIP connection. After the connections are in Asterisk, they don't require additional licenses. These calls only consist of an incoming VoIP portion and so only require one license.

✔ **Do you reinvite calls?** The G.729 license is only in use when a G.729 call is active on the channel. After a call is released by reinviting it to the far-end device, the license is disengaged and available for another call. This is yet another reason why reinviting is preferable. See Appendix B to find out how to reinvite calls.

After you determine the number of licenses you require, add 20 percent for a buffer and go buy them. Write down the analysis you have done to use as a template for the next time you evaluate your call volume.

If you're running more than one Asterisk server in a cluster, each server requires its own group of licenses. If you have two servers and 20 licenses, ensure that you purchase the licenses in groups of ten so that each server can get half the total. Loading all the licenses on one server in a cluster guarantees that G.729 calls to any other server in the group fail. Remember to spread the wealth.

The $10 fee gives you the licenses for as long as you need them. The only caveat to this is that the license can only be attributed to one server. However, you can reassign the license to a different server (for example, if the original machine dies, becomes lame, or is stolen). Just don't let anything happen to your second machine because you can only reregister the licenses once. If you have extenuating circumstances, contact Digium and ask to use the license again, but the decision is up to the folks at Digium.

We have found another Web site that sells the licenses from the company that owns the patent for the codec. Unfortunately, the licenses are geared toward very large users. The Web site (`www.spiro.com`, with pricing at `www.sipro.com/pdf/pdflp1.pdf`) lists the licenses for $1.45 each, but it requires a one-time buy-in of $15,000. This is a bit much for a small- or medium-sized Asterisk user.

Transmitting VoIP Faxes

Your standard fax machine is pretty boring. It may be able to scan and photo-copy, but it doesn't speak VoIP. This means that you have to treat it just like an analog phone if you want to transmit or receive fax transmissions using VoIP. The analog signal leaving your fax machine must be received by your Asterisk server on a port in an analog Zaptel card. After the signal passes that threshold, your dialplan converts the call to VoIP and directs it to the VoIP device of your choice.

Fax machines are very temperamental. They have difficulty handling VoIP transmissions that arrive out of sequence, are compressed, or are missing packets. If more than 3 percent of your packets are lost for any reason, your fax machine may rebel and lock up the transmission or end your call. If you send your faxes using the G.729 codec, you'll have this problem because it sends frequencies in the range of the human voice. Fax signals are in the extremely high and low frequencies, so they are ignored at times in the com-pression process of the G.729 codec. As a result, the compressed faxes fail.

Faxes successfully transmit over the G.711 voice codec. In spite of the fact that it's a voice codec and not a fax codec, the transmission still works at an acceptable level because the codec diligently converts all the sounds of the fax machines and transfers them to the remote fax. Because nothing is com-pressed, the packets are less likely to be lost or arrive out of order.

The downside to the G.711 codec is the incredible amount of bandwidth it uses. Many smart people have unraveled a way to maximize the bandwidth used on fax calls. You realize the same bandwidth savings as a G.729 voice call, but none of the information is lost in transmission.

This codec is called T.38. It doesn't attempt to compress the sounds used to transmit the fax; it treats the fax as an image. By converting the fax to an image file (such as a TIFF file), the information that represents the fax can then be transmitted to the receiving fax machine. The receiving machine is probably analog, so the last piece of VoIP hardware en route to the receiving fax machine gathers the data for the image and reconverts it into the squeals and screeches that fax machines understand.

T.38 is a newer codec. If you plan to use it, check with your VoIP carrier to confirm that it supports the codec. The process of implementing it into a network isn't that complex — it's simply time-consuming. Every VoIP switch in your carrier's network must be individually loaded with the software. Any switch without the software prevents the T.38 call from completing.

Asterisk can't convert between analog and T.38 at this time. It can, however, act as a "pass-through" for it. For example, a VoIP gateway at your inbound VoIP carrier can convert the fax to T.38 and send the fax through your Asterisk. You Asterisk merely sends the fax on to the final server that makes the conversion to analog and then back to T.38. Your Asterisk server can happily sit in the middle of the transmission and successfully relay the information for the call, as long as it doesn't have to interact with the conversion to or from T.38. The codec for T.38 in version 1.4 of Asterisk is identified as `udptl` and must be implemented to allow it to pass through T.38 traffic. Unless you're a local carrier using Asterisk as the backbone of your VoIP network, you probably won't be able to use this pass-through option.

Figure 9-1 demonstrates a pass-through transmission with Asterisk. The Real-Time Transport Protocol (RTP) stream is forwarded with Asterisk and isn't converted to fax transmissions until it reaches a VoIP gateway capable of converting T.38. Only after it reaches the T.38 gateway is the VoIP-based transmission converted to analog and sent on to the fax machine.

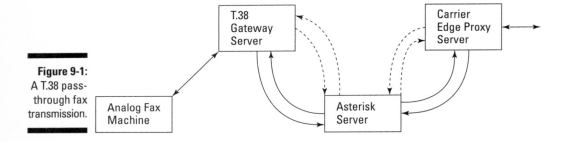

Figure 9-1:
A T.38 pass-through fax transmission.

We look forward to the day when Asterisk can convert T.38 to analog. This functionality opens a huge opportunity for Asterisk to function as a fax server to send, receive, and store faxes.

Receiving Videos

Asterisk also handles video transmissions. The codecs for video supported by Asterisk are as follows:

- ✔ H.261 utilizing 64 Kbps
- ✔ H.263 utilizing 20–30 Kbps
- ✔ H.263++
- ✔ H.264 utilizing 40 Kbps to 10 Mbps (in pass-through only)

Asterisk can forward video with compressed H.264, but it can't act as a gateway to convert the video to or from VoIP. Just like T.38 fax, Asterisk simply forwards the compressed information to the final VoIP device, which converts it back to video.

The current versions of Asterisk cannot reinvite the video stream, so you need sufficient bandwidth to send and receive the video. If you plan to supply high-definition video, expect each leg of the transmission to consume 40 Kbps to 10 Mbps. We are sure that the good people at Digium are working to allow Asterisk to reinvite the video stream in VoIP transmissions, but this feature hasn't been released yet.

Video codecs come in two varieties. The standard video codecs can play recorded movies, clips, or other prerecorded video. The second variety is designed to play real-time video streams. The codecs don't know whether the video they are being fed is live or recorded; they simply recognize it as a video stream and process it.

Unlike the audio and fax codecs, video codecs don't have many differences. Each new codec released is incrementally more efficient. The downside of the new codecs is that, well, they are new. It takes a while for a new codec to be adopted by the industry and applications to be upgraded to use the new bells and whistles. By the time the video applications respond and use the codec, it's outdated by another newer, faster, better video codec.

Video transmissions with Session Initiation Protocol (SIP) aren't unique. The only difference is that the video transmissions contain two extra RTP streams for the video in addition to the RTP streams used to transmit voice.

Video phone calls are a typical application of video on Asterisk. They require the installation of a Webcam and a softphone on your computer. You connect the video phone call just like a normal VoIP-to-VoIP call. As long as your softphone can send the video with the call, and the person you are contacting has a similar softphone that can receive and send the video, your video call is connected.

You can send a prerecorded video with SIP, but it isn't a standard VoIP application. This requires you to install a video capture card or connector device to allow the USB port on your server to connect to the RCA jacks on your video device. After you establish the video call, you can then play the movie from your video device. This is all very nonstandard, but it can be done.

Part III

Maintaining Your Phone Service with Asterisk

The 5th Wave   By Rich Tennant

"...so if you have a message for someone, you write it on a piece of paper and put it on their refrigerator with these magnets. It's just until we get our Asterisk sytem fixed."

In this part . . .

You find everything necessary to keep your Asterisk happy and healthy. In Chapter 10, you find out how to use the packet capture software Wireshark. It allows you to see the interaction between your server and your carrier on VoIP calls. This one small software program allows you to unequivocally identify who is rejecting a VoIP call and why.

Chapters 11, 12, and 13 take you into the nitty-gritty of troubleshooting. The vast majority of telecom issues you'll face come from outside your network. Keep this section bookmarked, because you'll refer to it for years to come.

In Chapters 14 and 15, we cover Asterisk call management and server environmental requirements.

Chapter 10

Troubleshooting VoIP Calls with Packet Captures

*V*oIP can be a frustrating protocol if you can't see what's happening inside it. VoIP makes you work under the assumption that the calls are being correctly sent and received from your hardware — which is a challenging situation to be in. If you order new phone numbers from your inbound VoIP carrier, you must have blind faith that the numbers are active and hitting your hardware at the prescribed time and day.

Asterisk includes too much ambiguity to comfortably run your company. If you are using Asterisk with VoIP, you can use a few tools to look under the hood at the transmission of calls. Peeking into the calls and seeing what is happening is much easier than contacting your carrier and waiting for someone to open a trouble ticket to perform the same type of call capture.

In this chapter, we show you how to use two of the best tools in VoIP to keep an eye on how your Asterisk is functioning: tcpdump and Wireshark, which is the new name for the open-source project formerly known as Ethereal. Almost all members of the initial Ethereal development team continue to work on the project, and we look forward to great things from this group of programmers.

Understanding Packet Capture Programs

Packet capture programs like tcpdump and Wireshark are specifically designed to capture packets of any protocol and display the information in an intelligent (or is that intelligible?) manner. They have the added benefit of breaking down standard protocols, such as Session Initiation Protocol (SIP), into their respective pieces so that the protocols are easy to read. This may sound like magic, but the programs only work well on text-based protocols. Fortunately, SIP and IAX — which are the protocols VoIP uses — are text based.

Packet capture programs aren't useful for analog or digital voice transmissions. The voice calls that use traditional telephony aren't comprised of packets, so the data that can be grabbed by a packet capture program isn't very useful. If you want to have some degree of visibility into traditional telephony calls, purchase a T-1 test set such as a Phoenix T-Berd 5575 or Fireberd. A new unit may cost $5,000 to $10,000, but you may be able to find an older refurbished model on eBay for a few hundred dollars.

Both software programs we write about in this chapter, tcpdump and Wireshark, have their pros and cons. Wireshark is a GUI-based system that allows for the easy distillation of packet captures. The filter and viewing options of Wireshark make it the best software to read any packet capture. You can also use the Wireshark GUI to initiate packet captures, but executing captures from the Asterisk Command Line Interface (CLI) is also very common. Because you can start both tcpdump and Wireshark from the Asterisk CLI, we prefer to use tcpdump to capture the information (we find that the coding required is simpler!).

Acquiring the Software

Wireshark can read the data received from either packet-capturing software. You can download Wireshark for free from the Wireshark Web site (www.wireshark.org). Remember to choose the version that matches the operating system of your computer or server. If you are using Wireshark to capture packets on your Asterisk server, download the Linux version. If you're using Wireshark to read existing packet captures from your desktop PC, download the Windows version.

Wireshark captures the entire IP packet data. It encompasses enough layers of information to allow you to interpret the actions taken by both ends of a transmission. Wireshark has the added benefit of understanding how to partition and display the packet information in a meaningful manner. Not only does it perform this task for the SIP VoIP protocol, but it also works with Real-Time Transport Protocol (RTP), Simple Mail Transfer Protocol (SMTP), Media Gateway Control Protocol (MGCP), IAX, and Domain Name System (DNS).

Wireshark allows you to read packet captures, even if Wireshark didn't execute the capture. The beauty of the software lies in its ability to exist on your Windows-based PC and read information captured with tcpdump from your Asterisk server running Linux. It doesn't matter what capture program gathered the data.

Another program geared toward Linux for capturing packets is tcpdump (www.tcpdump.org). It's also free, easy to download, and our preferred method of capturing data.

Installing the Packet Capture Software

It may seem like a simple procedure: Download the correct version of the software based on the operating system (Linux or Windows) and load it onto a server or PC.

We assume that your Linux server has access to the Internet and that is where you're retrieving the software. Wireshark and tcpdump each have complementary strengths, so we give you the code needed to install each. Capturing data with tcpdump is easier, but viewing the data is easier and more dynamic in its display with Wireshark. You can have both software systems available on the same server with no conflict. The software you install and use for a particular task — capturing or viewing — is up to you.

Figure 10-1 shows how to correctly position Wireshark in your network. You are installing packet capture software to validate the exact information being sent and received between your Asterisk server and VoIP carrier. Positioning your packet capture software on the server that contains the Asterisk software gives you the perfect vantage point to collect this data.

Figure 10-2 identifies the wrong place to install the packet capture software. Loading Wireshark or tcpdump on any server beyond the Asterisk server compromises the information you collect. Any conversation between the Asterisk server and your VoIP carrier is invisible to Wireshark or tcpdump in this position. Information can be gathered regarding the transmission of information from the Asterisk server to the media server on which Wireshark is loaded, but that doesn't validate the transmission of data between the Asterisk server and your VoIP carrier. If a DID number isn't loaded correctly and is failing, only a packet capture between the Asterisk server and your VoIP carrier can validate the source of the issue. You can see where the failure lies — whether your VoIP carrier sends the call that your Asterisk server fails to pick up, or your VoIP carrier fails to send the call at all — only if the capture is executed from the right piece of hardware.

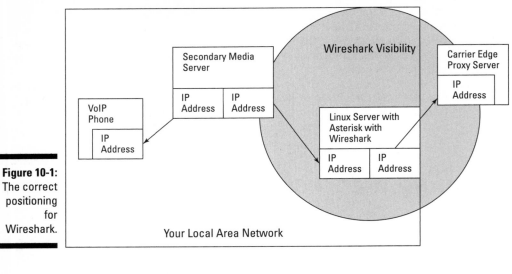

Figure 10-1:
The correct
positioning
for
Wireshark.

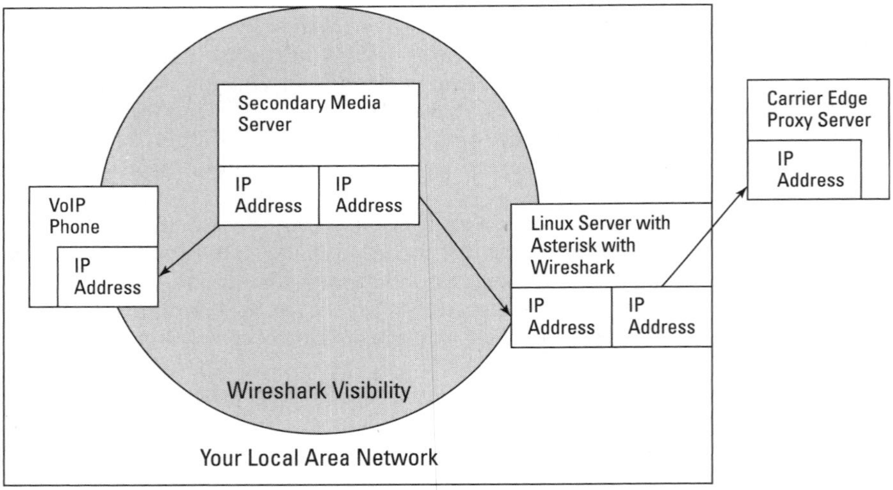

Figure 10-2:
The
incorrect
positioning
for
Wireshark.

Installing Wireshark

If you're using Wireshark to read the packet captures, you have to install the Windows version of Wireshark on your PC. The process of installing Wireshark on your desktop is just as easy as installing any Windows desktop application (such as WinZip or TextPad). *Remember:* You can't capture your server's VoIP information from your PC, but you can read and manipulate the data after the capture is complete. This is the easiest way to read and distill the packet captures. You can find a version of Wireshark for use in the Linux CLI (called *TShark*), but it doesn't have the GUI interface, so you're just as well off sticking with tcpdump.

If you're capturing packets with Wireshark, you have to install Wireshark on the server that is running your Asterisk software. You must install the packet capture software on the same server as your Asterisk software. Capturing data from any other location skews the information and renders the analysis flawed at best. The process of installing Wireshark onto the Linux server running your Asterisk is a bit complex, and having a background in Linux is helpful. Turn to Appendix C for these basics.

Dealing with Linux

All Linux packages aren't the same, so the installation process differs based on the distribution of Linux you're running. Despite the fact that the core operating system of each Linux distribution is similar, each distribution has inherent differences in how some tasks are executed.

Well over 100 distributions of Linux exist, each with an interesting name such as SuSE, MandrakeSoft, Ubuntu, MEPIS, KNOPPIX, and Damn Small. The main distribution Web site generally has ample information about the specific flavor of Linux you are using. Most of them have maintenance applications. The following table shows an extremely small sampling of Linux distributions and their maintenance applications.

Linux Distribution	Maintenance Application
Red Hat	Yum
Debian	apt-get
Gentoo	portage

The maintenance application may or may not be the appropriate command to install software. For example, the apt-get program in the Debian distribution of Linux also uses apt-get as the command to install software (in this case, Wireshark):

```
apt-get install wireshark
```

On the other hand, the Gentoo distribution doesn't use the name of its maintenance application as the install command:

```
emerge wireshark
```

Using a package management system, such as portage or apt-get, is the easiest way to bring the software onto your Linux server.

Execute the following commands from the Linux command-line interface (CLI) to install Wireshark on a Linux server:

```
mkdir /usr/src/tarballs
cd /usr/src/tarballs
wget http://www.wireshark.org/download/src/wireshark-
          0.99.4.tar.gz

gunzip wireshark-0.99.4.tar.gz

tar -xvf wireshark-0.99.4.tar

cd wireshark-0.99.4./configure
make
make install
```

Installing tcpdump

The `make` and `make install` commands ensure that the software is installed in the correct directory on your Linux server. You can install tcpdump at the Linux CLI with the following commands:

```
mkdir /usr/src/tarballs
(if directory does not exists)

cd /usr/src/tarballs
wget http://www.tcpdump.org/release/libpcap-0.9.4.tar.gz
wget http://www.tcpdump.org/release/tcpdump-3.9.4.tar.gz

gunzip libpcap-0.9.4.tar.gz
tar -xvf libpcap-0.9.4.tar
cd libpcap-0.9.4
./configure (possibly not needed)
make
make install
cd ../
gunzip tcpdump-3.9.4.tar.gz
tar -xvf tcpdump-3.9.4.tar
cd tcpdump-3.9.4
./configure (possibly not needed)
make
make install
```

Starting and Stopping a Packet Capture

We recommend capturing packets as efficiently and concisely as possible. Starting a capture and leaving it unattended results in the formation of a huge data file from which you must parse information. Start packet captures just prior to a test call, and end them immediately after terminating the call.

You can effectively capture data only on an interface located on the Linux server that's running the capture. This is why we recommend that you install Wireshark or tcpdump on the machine that contains your Asterisk software. You start a packet capture from the Linux console located in the root directory of your Linux server. The console for Linux looks just like the command prompt in DOS software.

The exact location of the packet capture command varies depending on how you loaded the software. A package management application may place it in one location, but you may have manually installed it somewhere else. Fortunately, you can easily track down the capture file by viewing the Linux command used to initiate it. The command commonly identifies the destination file location, allowing you to let the capture file reside in the default directory (where tcpdump was installed) or in another directory you've specified.

Capturing packets with tcpdump

The tcpdump command-line syntax can be quite complex and can include a lot of elements that you may never need to use. You start a packet capture from the Linux CLI with this command:

```
tcpdump -w newfilename.log expression
```

The command is made up of the following elements:

- ✔ `tcdump`: Tells Linux the software you are using.

- ✔ `-w newfilename.log`: Identifies the name that Linux attributes to the capture file. You can change the name for each packet capture, allowing you to keep each packet in sequence.

- ✔ `expression`: Identifies the specific elements to which you are attempting to restrict your search. Researching possible expressions from the manual page of tcpdump helps restrict your search to only the data you require. For example, the following expression captures only data on port 5060, to or from the specified host:

```
host 192.168.1.100 and port 5060
```

You stop a capture by pressing Ctrl+C.

Figure 10-3 identifies the participants of a VoIP call and the conversations that can be captured. The port or IP address you choose as the basis for your capture focuses the data received.

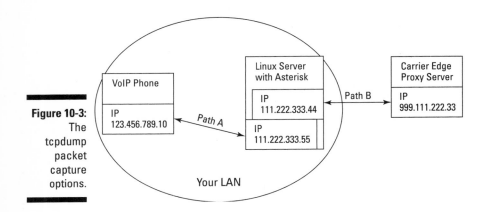

Figure 10-3:
The
tcpdump
packet
capture
options.

The most important element you need to capture data is the IP address or port from which (or to which) you are capturing the data. To initiate a capture from the main IP address of 111.222.333.44 listed in Figure 10-3, execute the following command from the Linux CLI:

```
tcpdump -w Newtestfile.log host 111.222.333.44
```

Packet-capturing software monitors a network interface such as an IP port in either a promiscuous or a nonpromiscuous mode. *Promiscuous mode* monitoring captures all packets traveling across the interface, regardless of whether they are coming from or going to your Asterisk server, or are destined for another computer on the same network as your Asterisk server. *Nonpromiscuous mode* monitoring restricts the packets captured to only the data intended to be received by or sent from the specific interface identified in the capture. Tcpdump runs in promiscuous mode by default.

The capture receives results in a file called Newtestfile.log that is saved in the same directory as the rest of your tcpdump software (unless you specified a full path name along with your filename). Because the default capture is done in promiscuous mode, the capture file includes information on all data transferred through that IP address.

To avoid distilling the information after the fact, simply execute the command in nonpromiscuous mode as follows:

```
tcpdump -w Newfilename.log -p host 111.222.333.44 or host
        999.111.222.33
```

Capturing packets with Wireshark

Capturing packets with Wireshark is not that challenging. You can start and stop a capture by following these steps:

1. **Choose Capture⇨Options from the main menu.**

   A window showing all your capture options pops up.

2. **Select options and confirm all choices in this window, including the interface, capture filter, and any other specifics you require.**

   The software provides drop-down lists whenever possible for interfaces and IP addresses on the server. The interface lists the network devices from which you can initiate the capture, and the filter refines the information.

3. **After setting your options, click the Start button to start the capture.**

   When you start the capture, a pop-up window that identifies the captured packets and running time of the capture appears. The window also contains a Stop button that you can use to terminate an active capture.

4. **Click Stop to stop the capture.**

You'll probably capture packets from the same interface and use one of a handful of different filters, so it won't be necessary to reset your options every time you initiate a new capture. After you have the options set as needed, you can run the packet captures by choosing Capture⇨Start from the main menu.

Figure 10-4 shows the standard Wireshark GUI interface. This screen shot is taken from a Windows-based installation of Wireshark, but the Linux version is similar.

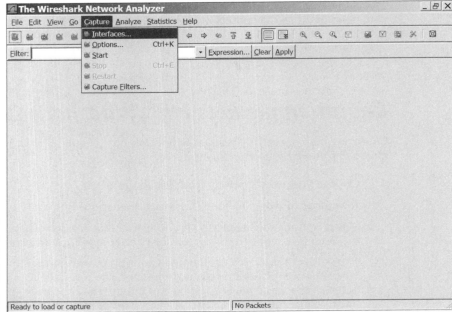

Figure 10-4:
The
Wireshark
GUI
interface.

Stopping the packet capture is done from this interface, and you can also name the capture file. While the capture is in progress, you can name the file by choosing File➪Save As from the main menu. You can do the same thing and bypass the menus by pressing Ctrl+Shift+S.

Reading a Packet Capture in Wireshark

Wireshark is our recommendation for viewing any packet capture, regardless of the software used to collect the data. You can open the capture by opening Wireshark first and selecting the file, or by double-clicking the capture file.

You can't do a quick sort in Wireshark to differentiate the packets you want to look at from the ones that were simply captured by accident. The easiest way to limit the data you must sort through is by starting the capture immediately before making a test call and ending it directly after the call terminates. If you are pulling the data from the main VoIP port used on all your calls, you may still have a large quantity of data to sift through.

If VoIP is a growing aspect of your business, develop your skills at reading SIP captures. The information contained in these captures can reduce a problem that may plague your company for months to something simple that can be resolved in minutes. Developing capture-deciphering skills is especially important if your core business is providing VoIP service to others. As your business evolves, you will encounter more issues both internally within your Asterisk and externally with carriers. SIP captures are helpful in determining whether a call failed coming into your system because you ran out of G.729 codec licenses or because it never reached your equipment from the carrier.

SIP captures have a degree of clout in the VoIP world. If your carrier asserts that a connection problem resides on your end, but the SIP capture proves otherwise, forward your capture to the carrier. E-mailing a capture and reviewing it with a company representative generally removes the anxiety and emotion of an issue. The captures don't tell you the specific source of a problem, but they can narrow the problem to a specific server.

Looking at a completed call

Figure 10-5 represents a simple call capture presented with Wireshark. You see the following three sections:

✔ The top section shows the highlights of the messaging and contains these columns:

- **No.:** Each line represents one call.

- **Time:** Shows the elapsed time from the start of the capture.

- **Source:** Identifies the source IP address of the call.

- **Destination:** Identifies the destination address that received the call.

- **Protocol:** Outlines the data protocol — such as SIP, RTP, IP, UDP, or TCP to name a few — used to write the data in the packet.

- **Info:** Shows the call's progression.

Select any line in the top section of the window to see the details of it in the bottom two sections.

✔ The second section provides expandable views for the specific elements of the transmission. Clicking the plus signs (+) to the left of the lines reveals more detailed information regarding the heading.

✔ The third section reveals the binary code for the line in question.

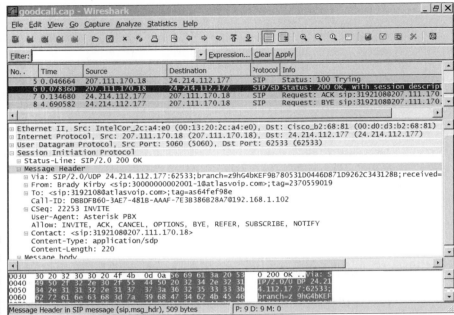

Figure 10-5:
A com-
pleted VoIP
call capture.

In Figure 10-5, our call was invited, authenticated, acknowledged, connected, and disconnected. We selected line item 6 for more in-depth research. We expanded the section regarding the SIP information, as well as the Message Header, Message Body, and the Session Description Protocol.

Looking at a failed call

Wireshark is a fabulous tool for troubleshooting. The benefits of the tool are obvious when looking at a failed call.

Figure 10-6 shows a capture of a failed call. All the information you need is listed in the top section of the window. Our Asterisk, at 24.214.112.177, attempted to call an SIP server at 207.111.170.18 and successfully authenticated, but was rejected. Everything seemed to be fine until we sent the INVITE requesting to be connected to the phone number 256-714-2943. The remote server attempted to make the connection but returned a 503 Service Unavailable message.

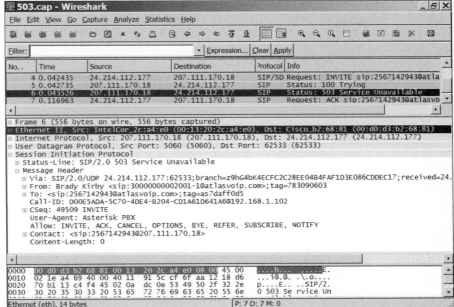

Figure 10-6:
A failed call
capture.

We could attempt further troubleshooting with Wireshark, but it isn't necessary in this case. We only need to identify the source of a failure or problem. The phone number we requested may be an invalid number, may not be programmed into the server we contacted, or may be knocked out by a natural disaster. The capture doesn't tell us the specifics but simply indicates the next company we must call to resolve the problem or receive details on the issue.

If the scenario was reversed and we were the company receiving the call at 207.111.170.18, the situation is different. The capture confirms that the call was sent to us, but we rejected it. This proves that the call didn't fail on the way to our server, but possibly failed at the originating VoIP server or was misdirected along the way. In this case, we have to track down the phone number and confirm that it is correctly established in our dialplan.

The following list details the most common SIP response codes you encounter when reading a capture in Wireshark:

✔ **100 Trying:** Notifies the originating server that the call is in progress after an INVITE is received.

✔ **180 Ringing:** Alerts the remote server to initiate ringing to the calling party.

- ✔ **183 Session Progress:** Is generally sent before a 200 OK packet to provide information about the remote called end of the call.

- ✔ **200 OK:** Indicates a successful call setup with this packet sent by the remote called party.

- ✔ **401 Unauthorized:** Used only by registrars. Proxies should use proxy authorization 407. Indicates a failed registration, generally due to a bad password.

- ✔ **403 Forbidden:** Accompanies calls restricted from connecting due to server configuration or dialplan setup at either the calling or called party site.

- ✔ **404 Not Found:** User not found. Indicates that the called party was not found on the server, and so the call failed.

- ✔ **500 Server Internal Error:** Signifies any unknown error on the server.

- ✔ **503 Service Unavailable:** Identifies a SIP server that is unreachable.

- ✔ **606 Not Acceptable:** Is generated when a user wishes to communicate, but can't adequately support the session required by the other end.

You can find a complete listing of the SIP response codes at the following Web site:

```
www.voip-info.org/wiki-SIP+response+codes
```

Chapter 11

Maintaining Your Telecom Services

*T*he one thing we can guarantee about the phone service connected to your Asterisk is that there will be a day when it won't work. The problem might be caused by an error in the code of your dialplan or a large network outage plaguing one of your carriers. The challenge is identifying the issue quickly and either resolving it, if it's within your Asterisk dialplan, or forwarding the specifics to the correct carrier, if it's outside your local-area network (LAN). As unlikely as it seems, a carrier might not know it has a large outage until you call in to report it.

This chapter covers general information needed to identify the source of an issue and describes the structure of the trouble-reporting network within your carriers. You may have three different carriers connected to your Asterisk, and providing the specific information each carrier needs helps to reduce your downtime. We also walk you through the basic information that telecom companies use to isolate problems, including a general overview of all the responsible parties that handle a call. Analog and digital calls are treated differently within carriers, so we cover the specifics of these before covering tips on managing and tracking your issues.

Using good troubleshooting etiquette

If you aren't used to dealing with phone problems, it can be a very stressful and frustrating process. These conditions sometimes make people aggressive and emotional, two traits that don't help when unraveling these issues with your carrier. As you go through the troubleshooting process, try to build goodwill with the carrier's representatives. They might run you through an automated system that takes two hours before you get to speak to a technician, but if the technician likes you when you finally do get to chat, the tech might let you him call directly the next time you have a problem. On the other hand, if you call the carrier with an attitude and venom in every word, the representative might be less than compassionate or concerned about the outcome of your problem.

Understanding Troubleshooting Basics

When you begin troubleshooting a problem, you are under the assumption that something in or connected to your Asterisk isn't performing at an acceptable level. For example, you might have echo when you call your grandmother or static when you call work. You might not be able to get through to your girlfriend because someone in the world of telecom thinks that her number has been disconnected or is no longer in service. The issue can affect all your calls, both inbound and outbound, or just outbound calls to a specific city. If you haven't modified your dialplan recently and the issue just surfaced, it's most likely not in the software of your Asterisk. Before you can begin testing a specific issue, you have to narrow the problem.

When troubleshooting an issue that has been identified as being a carrier issue, it's not your responsibility to correct the problem. Your main responsibility is to accurately report the issue. You can speed the process along by doing some troubleshooting yourself, but generally only your carriers can fix a problem. By isolating the issue to a single source, either a carrier or a piece of hardware, you can reduce the troubleshooting time by hours or possibly days.

No piece of equipment lasts forever. The process of using hardware heats the circuitry, ever so slightly, causing minor changes to how well it conducts electricity. Surges of electricity, power outages, overuse, and defective fans all expedite the process, but the cumulative effect is the eventual demise of the hardware. The challenge in telecom is identifying the one piece of hardware in the chain of devices handling your calls that has failed.

The good news is that your Asterisk is probably set up with different carriers for your VoIP service, your analog or digital phone lines, and your

InterAsterisk eXchange (IAX) connections — and possibly even another provider for the dedicated Internet line you use to connect to your VoIP carrier. This diversity of carriers allows you to compare them all, identifying the level of the problem.

Analog outbound long-distance call

An analog outbound long-distance call is one of the most basic calls people make every day. Your Asterisk transmits these calls from 1-, 2-, 4-, or 24-port analog Zaptel cards connected to regular phone lines, just like the one at your house. The call only qualifies as being long distance if it hits your long-distance carrier; that is, if the phone number you are calling is generally over 13 miles away. Figure 11-1 shows a long-distance call.

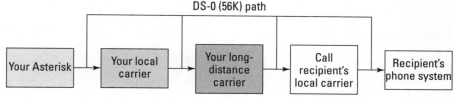

DS-0 (56K) path

Figure 11-1:
An analog outbound long-distance call.

Your Asterisk → Your local carrier → Your long-distance carrier → Call recipient's local carrier → Recipient's phone system

Calls flow in one direction.

The call starts with your Asterisk and then progresses from your local carrier to your long-distance carrier, to the local carrier that handles the service for the person you are calling, and finally to the phone you are calling. This is one of the most basic calls that you make and is probably the type of call that experiences the greatest quantity of issues.

Before you can pinpoint a problem to a variable in Figure 11-1, you need to compare the call to at least one of the remaining five call types.

The *Public Switched Telephone Network,* or *PSTN,* is a generic term for all long-distance carrier points of presence (POPs) and hardware that routes and directs phone calls, as well as the local carrier central offices (COs) and all their hardware that routes and directs phone calls. When a call is being sent through the PSTN, it is aggregated onto circuits with other calls for part of the path it takes from origin to termination. Imagine the PSTN as being a huge freeway for phone calls. All calls use the same roads, sometimes only until the nearest off-ramp and sometimes for hundreds of miles. If a problem on a call happens while it's being routed through the PSTN, the issue is said to be in the *switched network,* even if the call originally started or ended on a dedicated digital connection.

Analog inbound long-distance call

An analog inbound long-distance call is a very helpful call type because it eliminates your long-distance carrier as a potential cause of an issue. The long-distance portion of the PSTN is filled with hardware that routes and maintains the quality of your call. The greater the distance the call must travel through the PSTN, the more hardware is required to maintain the call. More hardware, of course, means more potential points of failure, so you are more likely to encounter a problem in the 28 or 2,800 miles where your long-distance carrier is handling your call.

Figure 11-2 demonstrates that when someone calls directly into your Asterisk (not using a toll-free number), the caller doesn't use your long-distance carrier. If you have a problem on both outbound long-distance calls and inbound long-distance calls, you can eliminate your long-distance carrier as the source.

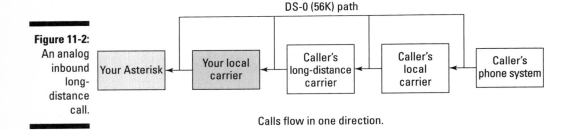

Figure 11-2:
An analog inbound long-distance call.

Calls flow in one direction.

Analog local call (inbound or outbound)

An analog local call uses the same carrier and interacts with the same hardware (carrier hardware and your Asterisk) for both outbound and inbound calls. It doesn't matter whether you call your friend across the street or whether someone calls you; the only carrier that generally ever sees your call is the local carrier you both use. Figure 11-3 shows this call type.

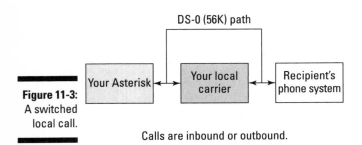

Figure 11-3:
A switched local call.

Calls are inbound or outbound.

The call barely enters the PSTN and might pass through only one central office. If you have a problem with a local call, the possible sources are your Asterisk and the local carrier.

Dialing someone else's toll-free number

When you dial someone else's toll-free number, the only piece of the call that you have any responsibility for is your phone system. Your local carrier is responsible for identifying the carrier that receives the traffic, and then your local carrier forwards the call to that network. Figure 11-4 shows the path the call takes when you dial someone else's toll-free number.

Figure 11-4:
Dialing someone else's toll-free number.

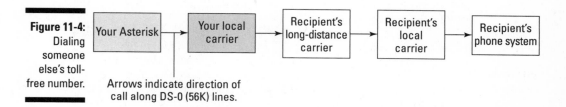

Arrows indicate direction of call along DS-0 (56K) lines.

If you have a problem calling someone else's toll-free number but all your other calls connect properly, the problem probably has something to do with either the carrier that handles the traffic of the toll-free number or the local carrier that terminates the call. We show you how to identify which carrier is the problem in Chapter 12.

Many companies use the same long-distance carriers. For example, if you use Sprint for your long-distance service and you dial the toll-free number of a company that also uses Sprint for long distance, you might see a similar problem on your outbound calls. In this case, you should focus on troubleshooting your outbound long-distance calls on the Sprint network. When Sprint resolves your outbound issue, the toll-free problem for the company you are dialing is probably also fixed.

Someone is dialing your analog toll-free number

The last analog call type is an inbound call to your toll-free number. The diagram for this call, shown in Figure 11-5, looks similar to that of the outbound long-distance call (refer to Figure 11-1). It includes the same path.

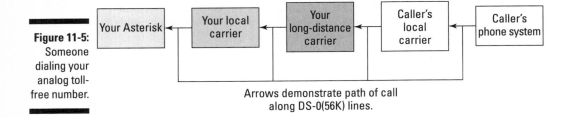

Figure 11-5:
Someone
dialing your
analog toll-
free number.

Arrows demonstrate path of call
along DS-0(56K) lines.

The main aspect that separates a toll-free issue from a regular outbound issue is the fact that the call is being sent to your long-distance carrier based on a database check done by the originating local carrier. Aside from this, the call encounters all the same carriers and hardware it would if you had made an outbound call to the origination number. This also includes the analog Zaptel card in your server.

Analog toll-free calls are processed by your Asterisk dialplan just as if they were direct-dialed inbound calls. After they enter your server, they're processed by the [inbound] context, just like a direct-dial inbound call. You might have special programming for the numbers based on the toll-free number dialed, but all incoming calls (toll-free or not) are handled the same way.

Digital calling

You can compare calls over your dedicated digital circuits with analog calls to eliminate your local carrier as a possible source of a problem. Figure 11-6 shows how your *local loop,* which is the digital wiring from your Asterisk to your long-distance carrier's network, passes through your local carrier but doesn't interact with it.

This is on a long-distance-carrier–provided dedicated digital circuit. Local-carrier–provided dedicated digital circuits terminate at the local carrier, and all calls enter the PSTN from that point.

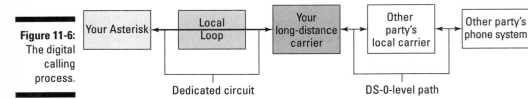

Figure 11-6:
The digital
calling
process.

Dedicated circuit

DS-0-level path

Only three elements, described as follows, interact at the individual channel (DS-0) level on a dedicated circuit:

✔ **Your Asterisk:** The digital Zaptel card receives the digital T-1 circuit cable into a phone jack (RJ45) and breaks it into 24 usable DS-0 circuits that transmit and receive your phone calls.

✔ **Your carrier's switch:** The one piece of hardware in your carrier's switch that you should be concerned with is the section that functions like your multiplexer.

✔ **Echo cancellers:** This hardware eliminate the echo on your phone calls. They may be in your long-distance carrier's network between the POP and the local carrier, or within the local carrier network.

The next important area to note in Figure 11-7 is the point from which the call enters your carrier's network to the point where it's delivered to the recipient's phone. Figure 11-7 shows an expanded view of a dedicated call.

The local loop begins at your building and ends at the Carrier Facilities Assignment (CFA) point, where the local loop enters your long-distance carrier's network within its POP. The majority of a dedicated call actually isn't dedicated but is handled by the PSTN, where it's routed and handled with every other call in your carrier's network.

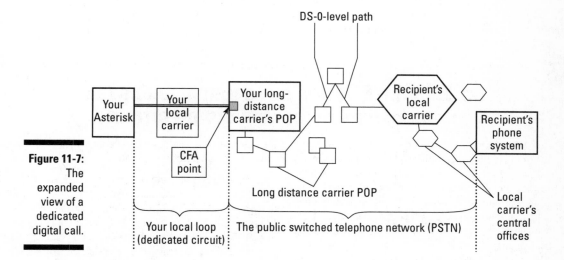

Figure 11-7: The expanded view of a dedicated digital call.

Connecting via VoIP

VoIP calls are the final type of call that can come into or go out of your Asterisk. The Internet circuit that connects your Asterisk server to the Internet makes this type of call similar to a dedicated digital call, but the troubleshooting varies a bit. The Internet Protocol allows a simple PING message to be sent from your Internet Service Provider (ISP) to your router to ensure continuity.

This is a simple cursory test to ensure that the public Internet can reach your router. In spite of the success of such a test, the digital circuit that provides the connection may have issues, and you may still need to test it intrusively to identify failing hardware or misoptioned equipment.

The ISP doesn't necessarily have to be the same as the carrier that provides the VoIP connectivity to the PSTN. You may even have one VoIP carrier for your inbound calls and a separate VoIP carrier for your outbound calls. As silly as this sounds, the requirements for Emergency 911 (E911) and directory listing service have steered some companies away from providing VoIP outbound calling service, so the pricing for each type of call may necessitate having separate VoIP providers. The good news is that you can at least connect via the same Internet connection.

Comparing call types

The call type diagrams in Figures 11-1 through 11-8 are great tools to use when you're troubleshooting. You should record the types of calls you have made and note whether they experienced the problem. Then you can refer to the diagrams and begin eliminating variables. For example, if you have static on local calls and inbound calls, but not on dedicated digital calls, you can distill the variables, as shown in Figure 11-9.

Keeping the PSTN in mind when you have a problem on a dedicated digital circuit

When reporting trouble on your dedicated digital circuit, remember that the local loop is the dedicated section your calls pass through. If you have constant echo, static, or dropped calls, or if the entire circuit is down, refer to your circuit when reporting the issue with your carrier. If you have problems you can duplicate by dialing out on an analog phone line over your carrier's network — such as calls failing to a certain area code, fax completion problems, static, or echo — open a trouble ticket with your carrier based on the call from your analog phone line. If the problem exists on both your analog and dedicated digital calls,

but you open the ticket as a dedicated issue, the problem can be harder to diagnose.

Carriers *do* assign greater priority to dedicated digital circuits, but they also push to intrusively test your circuit. The intrusive testing process prevents you from using any channel on your circuit during the testing. If you have the same problem dialing out from an analog line as you do from your dedicated digital circuit, report it as an analog line issue. This shortens the time to resolution by preventing your carrier from wasting time testing your dedicated local loop.

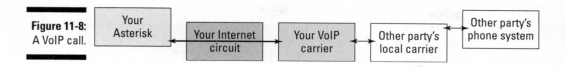

Figure 11-8:
A VoIP call.

When troubleshooting, look for similar paths. For example, when you compare the diagrams of the inbound calls and the local calls, you might see that the two calls only have two variables in common: your Asterisk server and the network of your local carrier. Local calls are commonly handled by your Asterisk server, your local carrier, and the phone system of whomever you are calling. If you simply call another person in your local area and have the same problem, you can eliminate the hardware of the first person you called as the potential problem. The only two variables left for you to investigate are your Asterisk server and your local carrier.

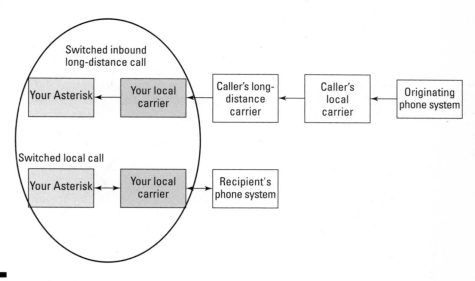

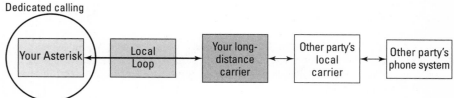

Figure 11-9:
Comparing the call types in diagram form.

Keeping an open mind about the information

The speed at which a problem is resolved is directly related to how well you accept the results of your testing. It's not uncommon for people to isolate an issue to a single variable but not troubleshoot that section first. This can occur when all the signs indicate that the issue is with your hardware — in your case, the Asterisk server. If your Asterisk server may be the issue, open a _technician assistance ticket_ (also called a _tech assist ticket_) with your carrier to schedule a time that a technician can walk you through testing and work with you. Your carrier's technicians can execute a series of tests that yield information to expedite the resolution of your problem.

Every inbound long-distance call coming from various companies in different sections of the world has its own local and long-distance carrier. Therefore, you can quickly eliminate the other long-distance carrier and the other phone system. The only variables that can be causing a problem are either your Asterisk or your local carrier.

After you consider the call from your dedicated circuit that did not have the problem, you can eliminate the internal routing in your Asterisk as a possible source of the issue. You could still have a problem with the specific analog Zaptel card, but because you have different ports configured for inbound and outbound calls, the problem may be contained to calls in one direction. If inbound and outbound analog calls have the issue, either your Zaptel card has failed or the problem exists within the network of your local phone carrier. You have now reduced the possible sources of the problem to just one entity (your local carrier). The next step is to call the local carrier and open a trouble ticket. It's helpful to let the local carrier know about the tests you have done and that the problem doesn't affect calls on your dedicated circuit, where they don't interact at the DS-0 level.

Narrowing carrier-level problems

If your problem persists on all your calls but you've narrowed the source to one of your carriers (eliminating your Asterisk server as a source), you can easily locate the issue and make sure that it gets repaired. If the problem doesn't affect all your calls, you might need to focus more on the pattern of failures within the carrier.

The following three additional questions can help save time when you open your trouble ticket:

✔ **Is the issue specific to a time of day?** Your carrier might be sending your calls over an overflow route that isn't very stable at peak network times between 10 a.m. and 3 p.m. If this is the case, the issue might be with a route that isn't frequently used or monitored. If the problem happens intermittently during your peak times and not during the peak times of your carrier, you might have a bottleneck in your Asterisk server. Time-of-day issues typically indicate a problem based on the volume of calls passing over a network, and identifying a problem this way can steer the technicians to a faster resolution. Share the information; it's good for everyone.

✔ **Is the issue geographically specific?** If you can call everywhere in America except the West Coast, the state of South Carolina, LATA 730, the 305 area code, or any geographic region that you can identify in a telecom-based group, note this information to your carrier. If a large outage has occurred affecting all calls to the West Coast, your carrier might not know this until you call.

✔ **Is the issue intermittent?** If the problem occurs randomly, regardless of time of day or geography, get ready to settle in for a longer troubleshooting process. With intermittent issues, first determine the percentage of calls affected. Issues that affect about 50 percent of your calls are not that difficult to find, because if your carrier watches ten test calls, the problem is bound to show up in around half of them, making it easier to research and repair. If the problem affects 5 or 10 percent of your calls, you may need to make 10 or 20 test calls before one call is affected. When less than 10 percent of your calls are affected by a problem, you enter the world of the needle in the haystack. As long as both you and your carrier stay focused and have the time to devote to troubleshooting, you will eventually find the problem. If the issue is low on your priority list, it might persist for weeks or months.

Creating a cut sheet

A *cut sheet* is a document containing all of the vital information about your phone service. The more complex your telecom inventory, the more you'll use your cut sheet to identify the services, account numbers, and contact numbers for every carrier you use.

Any delays in reporting trouble issues are undesirable, so create a cut sheet with the following information for each carrier you use. Print a copy for each employee who may need to call in to report trouble. The cut sheet should include the following items:

✔ **The name of the carrier:** This is just a nice header so that you know to whom all the other information belongs. Be sure to list your local phone carrier separately from your long-distance carrier and your ISP separately from your VoIP carrier.

- ✔ **Service provided:** You may have one carrier for your VoIP inbound service and another one for your VoIP outbound service. Calling the inbound VoIP carrier because you can't call out to someone delays resolution and makes you look disorganized. You may have redundant ISPs as well, so include one entry for each.

- ✔ **Account number/circuit ID/IP address:** Depending on the service with the problem, the carrier requires different information:

  - *Analog lines:* The carrier requires your account number and/or your phone number.

  - *Dedicated digital lines:* The carrier requires your circuit ID.

  - *Internet connection:* The carrier requires your circuit ID and the public IP addresses.

  - *VoIP providers:* Depending on the type of issue, the carrier may either need your account number or IP address if all your service is affected. If you have a single phone number that isn't working, the carrier needs a specific call example.

- ✔ **Trouble-reporting number:** List the primary phone number for the trouble-reporting department at each carrier as well as the escalation lists.

- ✔ **Phone number for provisioning department:** This may seem to be an odd thing to have on a cut sheet for trouble issues, but the cut sheet is a convenient place to keep this information. Adding or changing your service is easier if you don't have to search for the contact phone numbers to do so.

Some trouble issues are also identified as being provisioning issues. This is especially the case with VoIP carriers that migrate local phone numbers to their service. The *Local Number Portability (LNP)* process of negotiating the release from one carrier may result in the number failing for a period of time. Technically, the number is not failing because of a network issue but potentially because of a logistical issue. Always research LNP issues to determine if the number has actually been activated on your new carrier (a legitimate trouble issue), and not lost in the migration process to your new carrier (a provisioning problem). Carriers have a 24/7 staff for trouble issues and they are resolved quickly, whereas provisioning problems frequently take days to resolve.

The cut sheet is a good place to add other information. We recommend listing the name, phone number, and after-hours phone number for your hardware vendor. This information is handy in the event that the issue belongs to your server or other telecom-related hardware. Also, list all IP addresses of importance to you. This includes both your public IP addresses from which you send your VoIP traffic as well as the IP address of the VoIP carrier you connect to. If someone changes an IP address, it's easier to check it up-front than to wait hours for a technician to discover the change.

Getting the Most from Your Carrier's Troubleshooting Department

Most carriers have a two-tier structure for handling problems. The first tier of people are the entry-level customer service folks. These people generally work from a script that asks you specific questions to qualify your issue. After a customer service agent has all the information he or she needs, you're given a *trouble ticket number* for tracking purposes.

If the first tier of customer service can't resolve the problem, the trouble ticket is sent to the network technicians, who make up the next level of support. These people can manipulate the network, update switches, and perform tests, and they are empowered to fix the complex things that go wrong. These are the people you want to speak with when you have a complex issue.

If you have a difficult and intricate issue, give the first-level customer service people just enough information to open the ticket and then ask to chat with a technician. If you try to tell the customer service rep all the minutiae of the problem, it might get lost in translation from you to the service rep, to the notes, and to the tech. It's always safer to simply get the ticket open, ask for a technician, and then give a complete explanation to the technician.

Keep the following points in mind when chatting with the first-tier customer service staff:

- ✔ **The rep you talk to is probably working off a script that has required information. You can't proceed without going through this step.** If you don't have all the information you need, things won't go far, so be sure to gather as many details as you can before you call customer service. For example, if you have a problem dialing a phone number in Alaska, but you don't know the number you dialed because you lost your notes, the carrier can't open the trouble ticket.

- ✔ **All the information you give guides the way your problem is handled.** If you tell the customer care rep that your dedicated digital circuit can't dial to area code 414, the ticket is sent to the dedicated department to test your T-1 lines. If the same problem happens when dialing phone numbers in the 414 area code from your analog phone lines, the trouble ticket is sent to the department in your carrier that handles calls from the PSTN in the 414 area code to determine why that area code is failing.

You should always open a trouble ticket as an outbound switched problem if it affects that call type. This option is always the most direct route to resolution and prevents your carrier from being distracted by inconsequential aspects of the problem. The same holds true for an issue that affects both your inbound and outbound toll-free calls to a specific area. Opening a trouble ticket as a toll-free problem focuses your carrier on the toll-free aspect of the call (Systems Management

Server [SMS] database construction, areas of coverage, and nuances that are particular to toll-free numbers) and not on the overall network (which is actually the source of the problem).

✓ **The person who opens your ticket is one of your greatest allies.** He or she can escalate the ticket on your behalf, monitor its progress, and call in favors to resolve your issue. Because the customer service rep is capable of doing so much for you (and has the power to do nothing), always be nice. If the customer service person you are chatting with doesn't understand what you are saying, graciously ask to speak to a supervisor.

Identifying your call treatment

When you call to report a problem with your phone service, your carrier's rep asks you to describe it. The more specifically you can describe the issue, the easier it is to find and repair. The rep might directly ask you for the *call treatment,* or the symptoms of the failed or substandard call. This is where telecom can seem like a foreign language. You know what is happening, but putting it into words that can be understood might not be that simple. The main call treatments you can encounter are described in the following sections.

TECHNICAL STUFF

Understanding the translating and routing of calls

Every carrier that receives a call must determine where to send the call; then it must send it. These two activities represent two of the most common areas that cause the failure of calls. *Translation* is the process a carrier undertakes to identify the destination of the network that must receive the call. Long-distance networks don't send calls all the way to the end telephone that is receiving the call; the local carrier handles that job. The long-distance carrier sends the call to the local carrier's specified central office (CO). If the local carrier decides to use a different CO to ring your phone, but your long-distance carrier is not aware of the change, your long-distance carrier sends the call to the wrong CO and the call fails. The translation department at your long-distance carrier corrects the problem because the translation of the number you dialed to the correct CO must be resolved.

When a carrier's network determines where to terminate a call (by choosing the correct local carrier CO), the carrier has to deliver the call to that location. The *route* is the path the call takes from the moment it enters the carrier's network to the point it leaves the carrier's network. The routing department within each carrier monitors these routes and prevents them from being congested. If one main circuit fails, calls proceed through a secondary or tertiary route set up by the routing department. If a large outage occurs or calls fail on their way to the correct local carrier CO, the routing department identifies the fault in the network and repairs it.

Understanding why "your call cannot be completed as dialed"

If you hear a recording that tells you that your call can't be completed as dialed, you may have misdialed the number, the area code of the number you are dialing may have changed, or a translation problem may have occurred in a carrier network somewhere along the line. This recording generally indicates that the problem affects an individual phone number. Your ability to dial other numbers in the state, town, or country is usually not impacted. Try the number again, check the digits and the area code, and then make the call to your carrier's customer service if necessary.

Understanding why "the number you have called has been disconnected or is no longer in service"

A recording that says that the number has been disconnected or is no longer in service might be legitimate; the number has simply been disconnected. If you know that it's not disconnected, either you misdialed the number or a translation issue exists somewhere.

Handling an "all circuits are busy" message

On rare occasions, you may hear a recording that tells you that all circuits are busy. This recording rarely means that all the circuits available in your carrier's network are occupied, unless you are trying to call your mom at 9 a.m. on Mother's Day, along with everyone else in the world. This recording is generally played when your carrier has some type of outage and a portion of the network is down. If a backhoe accidentally cuts through a phone line, thereby taking down phone service for an entire city, you probably hear this recording when you try to dial out (the carrier unfortunately doesn't have a "Sorry, but our main phone line has been cut by a backhoe" recording).

Listening for tones and tags

Tones and *tags* are supplemental sounds or recordings and are generally attached to a standard recording. Listen for *tri-tones* that are played before recordings by a local carrier. They are generally three ascending notes that sound like they come from a cheap synthesizer — very shrill and high-pitched.

The tags are more important because they frequently list the switch that is playing the message. If you hear a recording of "your call cannot be completed as dialed — fifteen dash two," you know that switch 15 on your carrier's network probably has a problem. If you hear tones and tags during a recording, be sure to note them, because alerting a technician to the specific tags can shorten the amount of time it takes to solve your problem. Instead of tracking down the failed call, the technicians can go directly to switch 15 and analyze it. The number after the dash may correspond to the recording played (for example, the *two* in this example may mean "cannot be completed as dialed"). On the other hand, the number may have no significance. Every carrier has its own system for the tags played. The tags may mean nothing to some carriers, but to others, they can be significant. You should always have more information rather than less.

Understanding the fast busy signal

A fast busy signal is a busy signal that sounds twice as fast as the normal busy signal. You probably hear a fast busy signal when part of your carrier's network is down (the pesky backhoe again — see the section "Handling an 'all circuits are busy' message," earlier in this chapter), so your call can't be completed.

Handling echo, echo, echo

Usually, only one person on a call hears an echo. Carriers have specific hardware installed throughout their networks called *echo cancellers* (or *echo cans*) to eliminate echoes on calls. These devices can fail over time, be misoptioned, or be mistakenly installed backward. The interesting thing is that if one person hears an echo on a call, it's probably the result of a bad echo can on the other side of the call. Sometimes, both people can hear the echo, but it's not as common.

Echo isn't immediately visible to the technicians at your carrier. If your call fails to a fast busy signal, your carrier can pull the call record and find the switch that killed the call. Some issues, such as dropped calls or static, are visible in a circuit's performance report, which indicates that an electrical or protocol-related anomaly occurred. Echo on a call, however, doesn't leave a trail of breadcrumbs and is therefore difficult to isolate and repair.

The reverberation of voice that people hear and classify as echo might also be the result of something other than a failing or misoptioned echo can. This other type of echo is referred to as *hot levels* and is the result of the overamplification of a voice in the transmission of the call. If you didn't have an equalizer in your stereo and you cranked the volume up all the way, your sound quality would be horrible and you would experience *reverberation* (or what you might call *echo*). To compensate for volume loss as calls are transmitted across hundreds of miles over circuits within the network, the switches in the POPs boost the sound a fraction of a decibel — called *padding the call* — to keep the signal strong. However, if the carrier amplifies the sound too much, you can hear echo. What is worse is that *hot levels* have cumulative effects.

If a call travels 15 miles and is padded 0.4 decibels in the two pieces of hardware it encounters, you won't hear echo. In fact, as long as the call has fewer than 2 decibels of padding from start to finish, you won't have a problem. If the call is traveling from coast to coast and each of the six long-distance POPs pads the volume 0.4 decibels, the call now has 3.2 additional decibels in padding. In this case, you will likely hear an echo. The cumulative nature of hot levels makes them difficult to find and correct. In situations like this, the issue might persist for weeks before the carrier finds all the offending switches and reduces the padding.

Standing up to static

Only one side of a call usually hears static, but static affects both incoming and outgoing calls. It's generally caused by a piece of hardware slowly failing in the network. The static can be minor to begin with, but it grows over time until you can't hear the person you're calling, or the person can't hear you. As the issue evolves, the piece of hardware eventually fails and all your calls fail. Don't feel bad; the bigger the issue, the easier it is to find. When the static is noticeable on every call, it becomes simple to track down and replace the affected hardware.

Echo might be invisible to the technicians who are monitoring a network, but generally static isn't. Almost all circuits in the United States can be monitored for quality. This doesn't mean that someone from your phone company is listening in on calls to ensure that they sound clean, but computer files can capture unexpected electronic or protocol activity on a circuit. Static can be classified as *unexpected electric activity,* due to some device is experiencing an electrical short.The *performance monitors (PMs)* are diagnostic files for a circuit that record the electric and protocol anomalies. The challenge with static is finding the correct span that is causing the problem. If you're dealing with an intermittent issue affecting 5 percent of your calls somewhere in the switched network for your carrier, the possible locations of the failing hardware can seem endless.

Dealing with dead air

Dead air refers to the phenomenon that occurs when you hear nothing on the other end after you dial a phone number. You don't hear the dial tone anymore, but you also don't hear ringing; you just hear nothing. When you hear dead air, stay on the call for 30 to 60 seconds; you'll probably hear a fast busy signal if you wait long enough. Dead air is generally caused by a translation or routing problem that caused your call to be transferred to a piece of hardware or a circuit that no longer exists. Because the hardware no longer exists, nothing is there to send you a polite recording or busy signal. All you get is dead air.

Dead air isn't the same thing as *post-dial delay (PDD)*. PDD is the silence you hear for a few seconds before you hear the ringing. Every call has some PDD, although it might last only one second. International calls are notorious for long PDDs; 15 to 30 seconds may pass before you hear the phone ring on the other end.

Getting around clipping

Clipping is the technical opposite of echo. With echo, a voice is repeated; with clipping, sections of the voice transmission are lost. Instead of hearing the conversation in full, you hear only sections of each word, because a few milliseconds are lost every second. This problem is just like echo or hot levels, but instead of having too much volume on the call, you have insufficient volume. The troubleshooting process for clipping is the same as that for any quality issue and requires both patience and a large quantity of call examples.

Dancing around dropped calls

Phone calls that are disconnected before either person hangs up are deemed *dropped calls.* If your phone system loses power while you are talking, it drops your call. The same thing happens if you are calling over a dedicated circuit that suddenly fails or takes an electrical hit. Dropped calls are researched by your carrier, and the cause of the disconnection is identified by a *disconnect code* that is passed through the network when the call ends. The disconnect code identifies whether the call was dropped by the origination side of the call, the termination side of the call, or an unknown event in the carrier's network. Dropped calls are generally the result of a failing piece of hardware and typically become more frequent until the source of the problem is evident. If your carrier is causing the disconnections, its technicians can find the failing hardware by protocol failures in the protocol monitors of the circuit.

Handling aberrant recordings

Each carrier has a few standard recordings; you don't hear a ton of variety. Any other recordings you hear probably come from the default sound directory in your Asterisk server. Your carrier doesn't have recordings on file that say "We regret that you were unable to access an outside line" or "Your long-distance carrier is currently rejecting your call."

If you receive a recording that refers to your carrier, it probably was not made by the carrier's network. You might want to check with the carrier first to ensure that the message isn't in its playlist but then investigate the dialplan of your Asterisk.

Working around incomplete dialing sequences

If you dial a long-distance phone number and the call is dropped after you press six or seven of the digits, you have the following two possible sources to check:

- ✔ **Incomplete dialing sequence caused by local carrier:** If your local carrier perceives that the number you are dialing is a local call, it might try to complete it after the seventh digit. If the call is actually a long-distance one, you need to notify your local carrier immediately. Some local areas require you to dial all ten digits for a local call, but the rule still holds: If you can't complete the dialing sequence and you are positive it isn't caused by your Asterisk dialplan, it must be caused by your local carrier.

- ✔ **Incomplete dialing sequence caused by your Asterisk software:** This situation can also occur if your dialplan is only set to route ten digits and you're attempting to dial a longer international number. Because the international prefix of 011 takes up three digits and the country and city codes require an additional two to six digits, you can easily reach the limit set by your phone system. Your call is dropped even before it hits an [outbound] context. Check your dialplan to ensure that it allows 011* dialing.

If you are dialing from an analog phone line, you must be able to dial all the digits of a phone number before your local carrier processes your call, sending it to your long-distance carrier. If you have a dedicated circuit, check your Asterisk dialplan and then contact your long-distance carrier. The long-distance carrier can watch you dial the phone number and see every digit you enter. More importantly, the technician can determine whether the hangup on the call is being initiated by your Asterisk or by the network. When you know the cause of the disconnection, you know where to turn to resolve it.

Providing a call example

A *call example* contains detailed information that allows your carrier to follow the call's path from the moment you dialed the number to the point the call failed. As technical as the idea sounds, a call example is just the information you write down about a failed call. After you dial out and get a "cannot be completed as dialed" recording, dial the number again and write down the necessary information (see the following sections for more information about what to include in your notes). When the carrier finds the call's end point, the technician can begin correcting the issue.

Call examples function to not only tell the technicians where to look for the problem but also to allow the customer service rep to categorize the issue. Depending on the information you provide, the customer service rep sends your issue to a specific department for repair.

Call examples might not be easy to obtain in some instances. If you're calling a number you dial often and the call fails, you know all the information required to open a trouble ticket. The challenge occurs when customers dialing in to your toll-free number have an issue. Customers might not have your analog phone number to tell you that they couldn't get through and report the issue. Even if they do get through to you, it's not common to begin your conversation with a quiz about the specifics of a failed call attempt. As a result, you might have to ask one of your customers to make test calls for you. The specific information your carrier needs is listed in the following sections.

Call examples have a shelf life of about 24 hours. The specific information about how a call is routed is kept in your carrier's switches for a finite amount of time before it's overwritten with new, more recent call data. If an issue crops up on Friday at 5 p.m., you need to contact your carrier immediately. If you try to provide the call example from Friday when you come into the office on Monday, your carrier probably rejects it and asks you to give it a newer example.

The date and time of call

Every call is logged in to your carrier's switches by the origination time. If you made four calls to a phone number and only two of them failed, be sure to give your technician as much information as possible to help differentiate the completed calls from the failed ones. If you provide one call example and don't mention the other three attempts, the technician might find one of the completed calls and close the ticket because research indicated that the call didn't fail.

North America has four time zones, plus those for Alaska and Hawaii, so be sure to identify the time zone when you provide a call example. If you don't tell your carrier that the call was made at 8:00 a.m. Eastern Standard Time, the customer service person might record the time by using a different time zone.

The origination phone number

You might be one of 10,000 people who calls a specific phone number today. The only way to isolate your call from all the others is by referencing it to your phone number.

If your Asterisk has multiple outbound analog lines and you don't know which one originated your call, you probably won't know the originating phone number. This isn't a problem. Generally, the outbound phone lines you dial from are provided in sequence, so simply use any of the numbers as your origination and then tell your technician that you are dialing from a phone system. When the call examples are found, the technician can trace the call back to your office by matching the area code and the first three digits of your phone number.

The number you dialed and the call treatment

If you made 5,000 calls in the past three hours and one of them failed, your carrier needs to know which one of the 5,000 numbers you dialed was the problem child. Seems reasonable, yes? The phone number you dialed gives your carrier an idea of the geographic area you are dialing into and is essential for tracking down the problem.

See the section "Identifying your call treatment," earlier in this chapter, for more on what information you should include when you describe the call treatment.

Understanding when to provide multiple call examples

A phone call has many paths it can take to reach its destination. Depending on fluctuations in the capacity of the network between the two points, you could make 10 calls to the same number in 15 minutes and your calls may never take the same route twice. Understanding the complexities of the phone systems is the key to resolving intermittent issues. The larger a problem, the easier it is to track down and repair. If you hear dead air, static, or echo on 5 percent of all your calls, you need to provide as many call examples as possible.

Whenever you have an intermittent issue, it's helpful to provide to your carrier with clean-call examples in addition to problem calls. The technician can review all the calls and begin comparing the individual circuits the calls took. After the technician eliminates all the circuits on the clean calls and isolates any remaining similar circuits on the affected calls, he or she is more than halfway through troubleshooting your problem.

Managing Your Trouble Tickets

Many carriers are efficient, but the reality is that you are one of the tens, hundreds, thousands, or hundreds of thousands of customers the carrier has. It's also common to have a 2,000-to-1 ratio of customers to customer service reps at a carrier. If you have dedicated circuits and bill several thousand dollars a month, you might receive some added support, but the ratio is rarely much better for big-time customers than it is for individuals or small businesses. With this information in mind, you can easily understand that your trouble tickets get quicker resolution if you manage them yourself.

The format you use to manage your trouble tickets can be as complex and structured as you want. You can build a large relational database that tracks trouble tickets by carrier, issue type, start date, mean time to repair, and resolution, or you can go for something less structured. The least-technical way to manage your trouble tickets is to simply write down your call example and all the information you need to open the ticket on a piece of paper, and update it with times you called in and the status you were given. This information allows you to track how long the ticket has been open and to track its progress. Of course, you might want to type this information and save it electronically just in case it gets thrown away with the other scraps on your desk.

If telecom isn't your business, you might only open one trouble ticket a year, but if you expect to open one trouble ticket every month or week, consider creating your own trouble ticket form to organize all the pertinent information.

Understanding the timelines

Carriers have standard timelines for responding to problems, based on the severity of the issue. If you have one phone number in Minnesota that you can't reach, the carrier's internal policy may be to provide you with a callback on the status of the issue within four hours. If the DS-3 circuit is down and you have no phone service, the standard time for a callback might be two hours.

Carriers typically won't give you an escalation until they have failed to respond to you in the standard interval, although local carriers are much stricter on these timelines than long-distance carriers.

That doesn't mean that your carrier solves your problem in two to four hours; it simply means that you receive a callback from a technician in that amount of time. We recommend escalating trouble tickets if they don't experience progress. If you have additional information about a problem, you can always call back into your carrier and update the ticket with the new call example or results of a test. This is a good way to ask for updates without sounding pushy.

Coping with large outages

Every network in North America has had a substantial outage at some time. The source might be a fiber cut, a computer virus, or simply the growing pains of integrating a new technology like VoIP into the network's POPs. (By the way, a *fiber cut* is exactly what it sounds like — a cut cable causes a catastrophic loss of phone service.) When these problems hit, they are big, fast, and generally take at least five hours to fix. Software issues that clog a carrier's internal network can take from one hour to five days to fix and are difficult to pin down. The mercurial nature of software issues makes any estimated time to repair dodgy, at best. Your best bet in these scenarios is simply to call in every hour or so for updates. The story can and will change as it goes along, and you need to be informed about the progress.

A day in the life of a fiber cut

Fiber cuts are more tangible, so carriers can typically provide realistic timelines for repair. The greatest anxiety is waiting for the technicians to be dispatched and find the cut. If the cut occurs close to a large town, the technicians might be on-site in 30 minutes. If it's in a rural area, it might take three hours. After technicians are on-site, your carrier should be able to give you an estimate on the time to repair. For a simple fiber cut, the technicians have to trace the cables back from the split and dig to find an undamaged section of the fiber to splice in a new section of cable. When one side is connected to the new cable, the technicians then have to dig another trench on the other side of the cut to expose the fiber and then begin splicing the new section of cable into it. If you estimate one or two hours per side to dig each trench and splice the new cable, you should be in the right ballpark. Now, for example, if extenuating circumstances exist and the fiber cut is the result of flooding that eroded a hillside, which exposed and shredded a mile-long section of fiber, you can expect a much longer time for repair.

Troubleshooting International Calls

From your perspective, troubleshooting an international call is just like troubleshooting a domestic call. The only twist to troubleshooting international calls is that you might find some interesting similarities if you try your call over other carriers. We show you how to troubleshoot analog outbound issues in Chapter 12.

Call-treatment similarities exist because your long-distance carrier doesn't use its own network to complete calls to every country in the world. We can guarantee you that MCI, Sprint, and AT&T don't own all the cables and hardware and have a staff of technicians around the world to connect your calls into Senegal, Papua New Guinea, and India. Your long-distance provider uses an *underlying carrier,* a company specifically designed to deliver international calls from the United States to a specific country or region in the world. The interesting thing is that only so many underlying carriers provide service into each country, and every large domestic long-distance carrier probably has a service contract with every large underlying carrier. In other words, more than one long-distance carrier is using the same path to complete calls to Gifu, Japan, or Prague, Czech Republic, at any given time.

It's only possible to attempt your call over other long-distance carriers if you enable 1010 dialing in your Asterisk dialplan, or if your dialplan cascades from your dedicated digital long-distance circuit to your analog lines and then to your VoIP carrier. The easiest way to bypass your Asterisk and use a 1010+ dial around code is to connect a single-line phone to one of the analog lines connected to your Zaptel analog card. This is covered more in Chapter 12.

Your long-distance carrier can route your international calls over several underlying carriers. The choice of underlying carrier depends on the underlying carrier's completion ratios compared with all the other carrier choices at that time, as well as the price you are paying for your international calls. Some carriers have a premium group of underlying carriers available for international calls, but they can't place you on that group of carriers because they would lose money. If your business is focused on international calling, you might be better served by paying a few pennies more per minute for your calls, if you can realize a better call quality or completion rate. If you are opening more than one trouble ticket every few months on international issues, speak to your carrier about obtaining a better route.

Resolving International Fax Issues

International calls are prone to having quality and completion issues. You might have to wait 45 seconds before you hear the phone ring on the far end. When someone picks up the call, you might hear static, echo, or low volume on the call, making it difficult to hear what the other person is saying. You can still have a conversation, but it becomes a bit more challenging than you might want. It's not fun.

If a fax machine experiences any of these issues, the fax probably fails. Any issue that prevents the clean transmission of data in a timely manner typically kills a fax transmission.

International faxes can fail for the following reasons, and unfortunately, you can't do much to avoid these problems:

- ✔ **Antiquated networks in the destination country:** If you are trying to fax to a company in rural China or a small village in Africa, for example, the network might be old, outdated, and inherently prone to dropping calls. The company might be maxing out its bandwidth, and without enough bandwidth to allow the fax machines to sync up after the call connects, your fax fails. *Solution:* Make multiple attempts at a lower baud rate to complete your fax.

- ✔ **Receiving fax machine can't increase baud rate fast enough:** If your fax machine is transmitting at 64 Kbps and the receiving fax is 20 years old and has a maximum throughput of 1200 bps, your fax machine may not have enough time to slow down and the remote fax may not have enough time to speed up before your fax machine loses sync and drops the call. *Solution:* Reduce the speed of your fax machine and keep trying the call. Eventually, it should complete.

- ✔ **Fax machines time out before connection:** Every fax machine has a timer on it that disconnects a call if it's not answered. The factory default on this timer is about 30 seconds, and for domestic calls, that is acceptable. The problem is that international calls take more time to set up. An international call can experience a 30- to 45-second PDD. If the fax machine you are dialing to picks up on the second ring, it might be a full minute before your fax machine receives a connection. *Solution:* Set the wait-for-connect timer on your fax machine to 2 minutes (or at least to 90 seconds).

- ✔ **Carrier compression techniques on international calls:** Some underlying carriers work so hard to maximize their profits that they squeeze every call going over their networks to tenuous levels. This isn't much of a problem on voice calls, but fax machines need enough bandwidth to allow them to sync up after the call connects. If the call has too much latency, the fax machines can't sync and the call fails. *Solution:* You might have few options to resolve this problem, except to open trouble tickets and push your carrier to move the traffic to another underlying carrier.

Chapter 12

Addressing Call-Quality Concerns and Completion Issues

Your server running Asterisk potentially has analog, digital, InterAsterisk eXchange (IAX), and Internet connections that send and receive phone calls. Each type of interface brings its unique combination of testing options and limitations. After years of troubleshooting these issues on a daily basis, we've come up with some general and specific troubleshooting tips for these Asterisk connections.

Whereas the previous chapter focuses on the basics of telephony troubleshooting, this chapter shows you how to troubleshoot analog, VoIP, and IAX calls. We provide information to a very granular level in some instances to ensure that you can troubleshoot these issues with confidence. Digital interface issues are more complex, and we cover those in the next chapter.

Working through Analog Issues

Analog issues come through your 1-, 2-, 4-, and 24-port Zaptel cards configured to work with your Asterisk. (See Chapter 2 if you aren't sure how you've configured your analog card.) Analog issues are generally the easiest to troubleshoot because you have the tools within your grasp to identify the source of the issue.

On a grander scale, the basic tenants of analog troubleshooting are the basis of all troubleshooting. After a call from a dedicated digital circuit or a VoIP connection hits the long-distance network, it is routed and processed and can be troubleshot just like analog calls.

Confirming configuration and setup of analog service

Before troubleshooting call-completion or -quality issues on analog service, confirm that your Zaptel cards are correctly installed and configured. Start by reviewing the configuration of the channels by executing the following command from the Asterisk command-line interface (CLI) in Linux:

```
zap show channels
```

This command displays all Zaptel channels in your current configuration. If installed Zaptel cards or their channels are missing in the data, reestablish their configuration in the /etc/zaptel.conf file. Chapter 4 covers the specifics of how to install a Zaptel card.

Zaptel cards rarely fail. Usually the wrong driver is installed or the card wasn't configured in the first place. Be sure to validate the configuration on the card before handing over money for a new card.

After you confirm that the programming and configuration for the analog card are solid, you can proceed to in-depth troubleshooting. Regardless of whether the issue affects every line or a single channel, you troubleshoot it the same way.

The next section walks you through all the steps necessary to isolate and prove out every element in the handling of an analog phone call. We walk you through each company and piece of hardware processing the call, taking turns bypassing each piece one at a time to identify the source of the problem.

Step 1: Bypassing the Asterisk server

Unplug one of the offending phone lines from your Zaptel (or other) card and plug it into a standard analog, single-line phone. Try the call again and see whether it fails in the same way. If the call now completes and doesn't have static or echo, or you have dial tone where you didn't before, your Zaptel card has failed.

If you don't have a cheap, single-line phone, buy one. Keep it around the office for testing; you'll be surprised how often it comes in handy.

Step 2: Checking your long-distance carrier

When you've proven that the problem isn't in your hardware, the next step is to identify whether the problem is with your long-distance carrier — if you're attempting a long-distance call, that is.

If the call that's failing is to a recipient less than 13 miles away, your long-distance carrier probably doesn't touch it. If the call destination is farther than 13 miles away, it's a long-distance call.

The industry-standard test for validating your long-distance carrier is called a *700 test.* The 700 test got its name because you dial 1-700-555-4141 from your touch-tone phone (rotary-dial phones may not have access to the service). The 700 test, a free local call, is routed by your local carrier to a recording that identifies your interLATA long-distance carrier. If your long-distance carrier is AT&T, for example, the recording says, "Thank you for using AT&T." This test is almost always accurate, so investigate anything that doesn't sound correct.

If the test sends you to a recording that states "Thank you for using Sprint" or some other long-distance carrier you've never used, call your local carrier to have the problem corrected. If the test confirms your long-distance carrier, go to Step 3.

A *LATA* is a Local Access and Transport Area. It's a geographic area designed by the federal government to differentiate the responsibilities of local and long-distance carriers. Their size and shape don't have a rhyme or reason. Some LATAs encompass an entire state, whereas others look like they're the result of political gerrymandering. The only absolute regarding them is that any call that crosses a LATA border (an *interLATA call*) must be brought across by a long-distance carrier. Your local carrier can't complete a call from San Diego, CA, to Eureka, CA; it must hand the call off to a long-distance carrier for completion. Unfortunately, the mergers of local and long-distance phone companies have blurred the line of this rule.

Some competing local exchange carriers (commonly called *CLECs*), such as PAETEC and Mpower, might not provide the 700 testing feature but might instead send you to an automated voice that recites the phone number you are dialing from. In this case, the only way to determine where your local carrier is routing your long-distance calls is by calling your local carrier.

The 700 test only validates the long-distance carrier for an individual phone line. If you want to check multiple phone lines, you have to test them one at a time.

The 700 test only validates the *primary interexchange carrier (PIC)* that your local carrier has listed for your *inter*LATA long-distance traffic. Some local carriers offer a 700 test for your *intra*LATA PIC of 1-700-*your area code*-4141, but generally, you can only confirm the long-distance carrier for your interLATA calling.

Step 3: Bypassing your long-distance carrier

You can easily bypass your long-distance carrier. The network probably handles the call for the greatest physical distance, and as such includes more potential points of failure than any other carrier handling it. If your call completes when bypassing your long-distance carrier, you know that the problem is within its network. This test works regardless of the issue — whether it's static or echo, or your call failing to a recording of "the number has been disconnected or is no longer in service."

To bypass your long-distance carrier, retry your call over the same analog phone that's connected to the line that used to connect to your Zaptel card. Initiate the test call by dialing a 1010+ dial-around code for another long-distance carrier. For example, if your carrier is Sprint, dial the code for AT&T. Some dial-around codes are as follows:

- ✔ **AT&T:** 1010288
- ✔ **MCI:** 1010222
- ✔ **SPRINT:** 1010333

Redial the failed or affected call using the dial-around code. If you were trying to call someone in Milwaukee, Wisconsin at 414-555-4141 and the number was failing, you would redial the number over the AT&T network by dialing the following number:

    1010288+1+4145554141

If the call completes, the problem resides within your long-distance carrier's network. Call your carrier and open a trouble ticket. Be sure to tell the carrier that the call completes over AT&T (or whatever carrier you used for the dial-around call). This helps to focus the troubleshooting and resolves the issue faster.

How dial-around codes work

A dial-around code tells your local carrier to immediately route your call onto the long-distance network that's associated with that particular code. Dialing a phone number with one of these 1010 codes bypasses the local carrier process that determines whether your call is local or long distance and identifies which carrier is assigned to your phone line for long distance. Your local carrier simply sends your call to the network attributed to that dial-around code.

If the call still fails when you use a different long-distance carrier, move on to Step 4 to prove out your local carrier.

Because you don't have a contract with the carrier that you used in the dial-around test, it will likely charge you the highest rate possible for the call. If you must make the call, expect to pay upward of 50 cents per minute with the possibility of a connection fee of $2.75 or more. You're charged only if the call completes, but don't be surprised to see a $35.00 charge on your phone bill for a 20-minute call. If you are dialing internationally, the rates could be several dollars per minute.

Step 4: Using a different local carrier

The last test involves checking whether your local carrier is the problem. Every call is processed through the local carrier, which identifies whether the call is local, long-distance, or toll-free before sending it downstream to the correct carrier. Some issues are easy to attribute to your local carrier. If you experience any of the following issues from the phone line you have removed from your Zaptel port and connected to your single-line phone, you have a local carrier issue:

✔ **No dial tone:** Your local carrier provides the dial tone for your calls. If any line still has dial tone, simply dial 611 from it (if you have a standard local carrier) and report the problem. If none of your lines have dial tone, call the toll-free number for trouble repair from your cellphone.

If you are listening to the line from a phone connected to your Asterisk, the dial tone you hear comes from the Asterisk. This is why you must unplug one of the incoming phone lines that terminate in your Zaptel card and connect it to the single-line phone to confirm whether the dial tone is actually reaching you.

✔ **Calls failing before you complete the dialing sequence:** If you are dialing a 10-digit long-distance number from the single-line phone plugged directly into the wall (bypassing your Asterisk) and the call fails before you can finish dialing all ten digits, your local carrier is mishandling the call. Your local carrier processes all ten digits before it forwards the call to your long-distance carrier. The call failing before the carrier can determine where you are calling indicates a problem within your local carrier's routing.

✔ **Static on the line before you finish dialing:** If you hear static or echo before the call is handed off to your long-distance carrier, the line-quality issue resides within the network of your local phone carrier. Listen carefully to the call to determine whether the static or echo begins before you input the last digit of the phone number you are dialing. If the call is clean until you finish dialing the number, your local carrier is most likely not causing the problem.

If none of these issues apply, bypass your local carrier by attempting the call from your cellphone (if a different local carrier provides service to your cellphone). If the call completes over your cellphone, your local carrier is the source of the problem.

Confirming problems on the receiving end

If you have the same failure or call-quality issue when dialing over your cellphone, and over another long-distance carrier, the problem is most likely caused by the local carrier at the far end of the call. In the telecom world, all roads lead to the final local phone carrier. Every phone number is owned by a local phone carrier that is responsible for providing dial tone and local service to you. Specifically, the local carrier central office that resides closest to your home or business provides your service. That central office is also where all calls into your phone number must pass through before they are sent down the line to ring your phone.

If the central office has a problem, the phone number you're dialing can't receive calls, regardless of which local or long-distance carrier handles the beginning of the call. If you are calling a number that fails regardless of the 1010+ dial-around code you use, or which local carrier you try, the local carrier at the far end has the problem. You can't do much to resolve this type of problem other than wait. The person you are trying to call probably knows that no one can call, and hopefully the carrier is working on it. Wait a day or two and try making your call again.

Keeping your local carrier from passing the buck

When you tell your local carrier that you have problems dialing a long-distance number, the customer service rep will quickly try to end the conversation and direct you to your long-distance provider. Typically, the local carrier's rep is right to direct you to the long-distance carrier.

You can easily clarify the situation by telling the person that you dialed the number with the dial-around code and it completed, but that when

you dial it without the dial-around code, the call fails. The customer service rep then knows that you have tested the issue. As long as you continue to press this point, the rep has to open a ticket for you and repair the issue. If you want to really impress the rep, tell him or her you completed a 700 test before you used the dial-around code.

Getting Analog Toll-Free Troubleshooting Basics

Toll-free troubleshooting is slightly more complex than standard outbound troubleshooting because the carrier that is receiving calls on a toll-free number can vary by state, LATA, area code, or area code and the next three digits in the phone number. You also have dual concerns of what the RespOrg is (see the next paragraph) and which one carries the traffic. Finally, a problem with toll-free numbers might not initiate from the toll-free number, but can be a problem with the regular phone number it rings into or just be a systemic network issue that affects all calls that run through the point of failure.

RespOrg stands for *Responsible Organization* with regard to toll-free numbers. The RespOrg of a toll-free number is the business entity that controls the toll-free number as recognized by the national toll-free database called the Service Management System (SMS) database. This company alone is enabled to redirect the traffic for the toll-free number to its carrier of choice.

Before you begin troubleshooting your toll-free number, you need to validate the toll-free number's basic information, as follows:

✔ **The toll-free number or feature:** Has it been working in the past? If you are in the process of moving your toll-free number from one carrier to another, or adding a new service, you can have any number of provisioning problems that prevent the service from being activated. If the service has never been established, call your carrier and review the problem with it. The number or feature may have never been fully activated.

✔ **The ring-to number:** The analog phone line that receives the toll-free call is referred to as the *ring-to number*. If you ordered the toll-free number to ring to your fax machine rather than the main phone line for your customer service department, that might explain why everyone who calls receives an unpleasant squeal in their ear. Check the phone number to which the toll-free number points. If your order is wrong, that could the source of the problem. If nothing else, you need to know the ring-to number of your toll-free number for the testing process.

✔ **The area of coverage:** If you only ordered coverage for the contiguous U.S. 48 states and calls from Alaska and Hawaii fail, the source of your problem is that you didn't request access from those areas. Coverage on toll-free numbers can be delivered in many different levels, and your available options depend on what your carrier provides. Some carriers offer many options that include or exclude Canada, Alaska, Hawaii, Guam, U.S. Virgin Islands, and/or Puerto Rico, whereas other carriers give access to all these options by default.

You can troubleshoot any issues, either quality- or completion-related, with the steps in the following sections. If you have quality issues, gather as many call examples as possible, both affected calls and clean calls, so that your carrier can isolate the issue. Completion issues are much easier to find, because they have a definite failure point that technicians can duplicate and find.

Step 1: Dialing the number yourself

When a customer calls in to your direct phone number to report that your toll-free number is failing, first find out the following information from your customer:

✔ The phone number and location from which your customer is calling

✔ The time of the customer's failed call

✔ The call treatment

Then try to dial the number yourself. It's more common for a toll-free number to fail from all locations than from only a specific geographic area, such as a LATA or a state. If your call fails when you dial the toll-free number, all inbound calls are probably failing, regardless of the location of the caller.

If you know the analog phone line that belongs to the ring-to number, unplug it from your Zaptel card and plug it into your single-line phone (just like you did in the test for outbound calls). This eliminates the potential that either your Zaptel card or your dialplan is causing the failure.

If you don't have a single-line phone available, check the Asterisk console to confirm that your server is receiving the call. You must either be a Linux root user or a user with permissions to execute the following command:

```
asterisk - r
```

If your call to your toll-free number completes, the problem might be restricted to the area your customer is calling from, which is why you need to know the phone number and location of the caller. You can ask the customer to help you troubleshoot the number, or you can call your toll-free carrier to open a trouble ticket.

You should also ask your customer to place you on hold so that he or she can try the toll-free number again. The customer could have dialed the number incorrectly. If the customer can complete the call, he or she probably misdialed the number the first time. If the call fails again, the problem could be geographical in nature. Until you receive additional reports of problems from your customers, or you get feedback from your carrier, you won't know how large the affected area is.

Write down all your call examples and call your carrier. When you open your trouble ticket, be sure to write down the date and time as well as the customer service rep's name. After the ticket is open, follow up on it every two to four hours until the issue is resolved.

If the call fails when you dial your own toll-free number, the outage might be affecting everyone who is calling, but at least you can troubleshoot the issue. In this case, clear a page in your notebook for test calls and proceed to Step 2.

Always try to find someone in an affected area to retest to your toll-free number. Long-standing customers or vendors are generally happy to spend five minutes making test calls. If your toll-free number always completes when you call it, don't assume that the issue is fixed until you receive a call from the area that was previously blocked.

Step 2: Dialing the ring-to number locally

Calling the ring-to number directly can identify whether the problem is within the toll-free routing portion of the call or is caused by the local carrier having problems reaching your regular phone line. Figure 12-1 shows the area of your local carrier's network that you're validating.

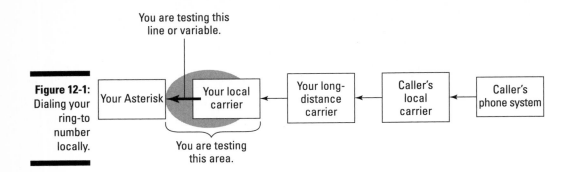

Figure 12-1: Dialing your ring-to number locally.

If you are calling from the same office that receives the toll-free number, you placed a local call to validate the ring-to number. Local calls pass through your phone system and your local carrier, so if you can complete this call, you know these two portions of the system are working.

If the local call to the ring-to number fails in Step 2

If your local call fails, the issue has nothing to do with the toll-free number. You can't complete a toll-free call if your local carrier can't complete its portion of the inbound call. Because you've already eliminated your phone system as a potential source of the problem by receiving the call on the single-line phone you installed on the analog line, the only potential source of the problem is your local phone carrier.

Collect all the information about your inbound local call examples, and dial 611 to reach your local carrier to open a trouble ticket. Write down the trouble ticket number, the person whom you spoke with, and the time and date, and then follow up periodically until the problem is resolved.

Don't mention anything about your toll-free number to your local carrier when you open your trouble ticket. If you try to explain that you have a toll-free number that isn't working and that you have been troubleshooting it, the local carrier can identify the number's RespOrg and direct you to call your long-distance carrier. Instead, open the trouble ticket on a call example that shows that you can't receive a local inbound call. That way, the carrier knows the issue exists within its own network. As long as your local carrier isn't confused by stories of toll-free numbers, the carrier can repair the issue in short order.

If your call completes in Step 2

If your call fails when you dial the toll-free number and completes when you dial the ring-to number, you can rest assured that the source of the problem is neither your local carrier or your Asterisk server. To resolve the issue, move on to Step 3.

Step 3: Dialing the ring-to number through your long-distance carrier

Your next step is to determine whether the issue is legitimately in the toll-free handling of the call or whether it's simply a larger network issue within your long-distance carrier. To truly validate how your long-distance carrier is sending calls to your phone number, you need to have a call placed over the long-distance carrier's network to your ring-to number. Figure 12-2 shows how the call is processed.

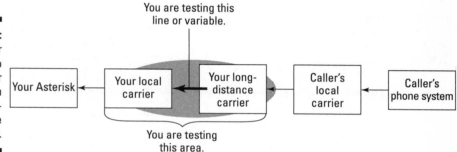

Figure 12-2: Dialing your ring-to number through your long-distance carrier.

You are testing this line or variable.

| Your Asterisk | ← | Your local carrier | ← | Your long-distance carrier | ← | Caller's local carrier | ← | Caller's phone system |

You are testing this area.

If your local carrier provides your toll-free number, that carrier has a complementary carrier that provides the long-distance service. In this case, you need to know the 1010 dial-around code for your long-distance carrier so that you can make the test call.

Test-calling the ring-to number on your long-distance carrier's network

Even though calling from one line in your office to another line in the same office is a local call, you can force the call to be routed over your long-distance carrier and back in through your local carrier.

Make this test call from the single-line phone connected to an outbound analog line. If you just finished the test call into the single-line phone on the analog line that receives your toll-free calls, don't forget to plug the inbound line back into your Zaptel card and plug the single-line phone into an outbound analog line. Trying to call your analog toll-free number from the number that receives the call only gives you a busy signal.

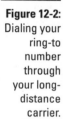

Understanding inbound routing

The most important piece of information in the world of telecom isn't the physical location of the phone, but the physical location of the central office that provides your dial tone from the local carrier. Every call you receive is sent to that central office and forwarded to you.

Periodically, your local carrier might decide that it has too many calls running through one central office, so it moves a group of phone numbers to a different central office in the area. The local carrier then updates the national Local Exchange Routing Guide, or LERG, database to inform every telecom company about the change. However, if your long-distance carrier is slow to implement the change, it might continue to send calls to the old central office, causing the calls to fail.

If your carrier is handling your toll-free calls, those calls also fail. The repair can take as little as 15 minutes if your carrier can identify the issue, or it can drag on for days. If you are aware of any changes your local carrier has made to your service, you should let your carrier know because it can reduce your repair time.

The only way you can dial a 1010 dial-around call through your Asterisk is by building it into your dialplan. It isn't difficult and can be done. But if you use the single-line phone and bypass your Asterisk, you remove it as a potential source of a toll-free problem.

By prefacing your call with the 1010 dial-around code of your long-distance carrier, the call is sent from your local carrier to your long-distance carrier and then forwarded back to your local carrier to ring to your office. It's a short loop, but even a short loop can indicate the general health of your long-distance network. If your long-distance carrier is Sprint, dial your phone number like this:

```
1010333 + 1 + area code + your phone number
```

Use 1010288 if your carrier is AT&T; use 1010222 if your carrier is MCI.

If the test call completes in Step 3

If your call to the ring-to number using your long-distance carrier completes without a problem, you have validated that at least a portion of the switched network is working. Carriers commonly route toll-free calls differently than direct-dial calls. Regardless, you must accomplish your next level of research with the help of the customer service reps at your long-distance carrier, so jump to Step 5.

If the test call in Step 3 fails

If your call to the ring-to number over your long-distance carrier fails, you might have an issue with your long-distance carrier's network. The good news is that your toll-free number isn't the source of the problem. The bad news is that your long-distance carrier hasn't updated the latest LERG database and is routing your calls to the incorrect central office (CO) of your local carrier (if you are having call-completion issues), or the carrier is sending the call over a defective route path (if you have quality issues). Next you need to determine whether any long-distance carrier can complete a phone call to you. To find out, move on to Step 4.

Step 4: Dialing the ring-to number over another carrier

So far, if you've been following all the preceding steps, you have isolated the problem with your toll-free number to the path your call is taking through your long-distance carrier to your ring-to number. If your local carrier is servicing your phone number out of a different central office, it's possible that only your local carrier knows where to find you. Figure 12-3 shows the area of a toll-free call that you are isolating by performing the test in Step 4.

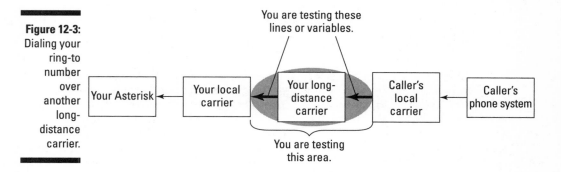

Figure 12-3: Dialing your ring-to number over another long-distance carrier.

Use a different long-distance carrier to validate the inbound route path. That is, if you use Sprint, you might use the dial-around number for AT&T. In this case, the test call would look like this:

```
1010288 + 1 + area code + your number
```

If the call completes, AT&T charges you for the call, at any per-minute rate it chooses. If the call doesn't connect, you won't be charged for the call. If your test call completes, keep the conversation short to prevent being shocked when your phone bill arrives.

If the call in Step 4 fails

Try a few other long-distance carriers, such as MCI (1010222) and Qwest (1010432). If these calls fail as well, your local carrier has hidden the inbound path to your phone from all long-distance carriers. In this case, take all the call examples for the multiple carriers you have tried and dial 611 to open a trouble ticket with your local carrier. Just as with all trouble tickets, you need to write down the trouble ticket number along with your call examples and follow up every few hours.

If the call in Step 4 completes

If the test call to your ring-to number completes over another carrier, your testing is over. This test only changes the long-distance carrier used, and every other variable remains the same. Because the results changed when you changed the long-distance carrier, the problem has to be with that carrier.

When speaking to your carrier, open your trouble ticket as an outbound call issue and, if possible, *don't* mention your toll-free number. Instead of telling the carrier that someone unsuccessfully tried to dial in on your toll-free number, tell the carrier that someone with the same long-distance carrier tried to call you and failed, so you did a test call that failed. If you begin talking about your toll-free number, the carrier might open the trouble ticket on the toll-free number instead, which can delay the solution.

When you open the trouble ticket, be sure to tell the technician that you forced the call from your office to the other phone number in your office by dialing the 1010 dial-around code. Then, as usual, write down the trouble ticket number, the name of the person you spoke with, and the time you called. Also make note of any additional testing that was attempted and changes made to the toll-free number or the network. Then set up a schedule to call back every few hours for updates.

Step 5: Validating the ring-to number and RespOrg

If you end up at this step, you have eliminated all the other steps that might be causing this problem, except for the processes and mechanisms that make toll-free routing unique to outbound calling. You should also log some failed call examples to your toll-free number and some completed call examples to your ring-to number.

After you reach a customer service representative, you need to validate the following two additional pieces of information before you open a trouble ticket:

- ✔ **Confirm the ring-to number.** This number might have been updated or input incorrectly. If your area code is 206 and the ring-to number is going to area code 602, you have found your problem. In about 15 minutes, your carrier can point your toll-free number where it needs to go and it will be working fine.

- ✔ **Confirm the RespOrg.** It's uncommon for a working number to suddenly be moved from one long-distance carrier to another, but it does happen. If your toll-free number somehow migrated to another carrier by accident, someone at the new carrier might have later realized it didn't belong to that carrier and blocked it. If the RespOrg isn't your carrier, you have to begin the process to move it back. We recommend taking two aspirin and then calling your carrier for assistance and the required paperwork. Unfortunately, your problem is now considered a *provisioning issue* and not a *trouble reporting issue*. This means that carriers deal with it with less urgency; the time to repair can be 7 to 10 days as opposed to 15 minutes or a few hours.

This is the one time when screaming might pay off, as long as you are screaming at people who can actually help you and as long as they are people you never expect to speak to again. We're talking about the conversation you need to have with the carrier that is somehow now in RespOrg control of your toll-free number. If you add emotional urgency to your discussion with the representative you speak to, he or she might not hide behind the normal 7- to 10-day time frame for migrations. Before you have this emotionally charged conversation, however, chat with your carrier first to make sure that the carrier is not at fault for letting go of the number.

If the ring-to number and the RespOrg match, proceed with the trouble ticket. Make note of the trouble ticket number, the person you spoke with, and the time and date you chatted (in case the problem takes more than one day to repair), and make some brief notes that you have checked RespOrg and the ring-to number. Then, set your alarm clock to ring every two or three hours so that you can follow up and keep the ticket moving forward.

Distilling the Nature of VoIP Issues

VoIP is a fascinating hybrid. It has a dedicated Internet connection that behaves similarly to a regular dedicated T1 voice circuit while your calls emerge from it into your carrier's network traveling with calls from analog lines. On top of these generalities, you have to add the complexities only possible with VoIP. You find latency issues, Real-Time Transport Protocol (RTP) issues, and outbound and inbound calling issues hovering around new or migrated phone numbers. These all make for a complex environment where you can easily be overwhelmed.

The first question to answer when troubleshooting VoIP is whether the problem affects all the calls on your Internet circuit or just select calls. If you can't send or receive calls that pass over your Internet circuit, but your calls passing through the analog or digital Zaptel cards are working fine, you have an Internet connection issue. Call your Internet Service Provider and report the trouble so that the technician can identify, isolate, and resolve your problem.

You may have a different VoIP carrier for your inbound traffic than you do for your outbound traffic. Identify which service is affected and open a trouble ticket accordingly.

Addressing VoIP call-quality issues

Call-quality issues are a bit trickier with VoIP than with traditional telephony. The fact that the voice portion of the call is separate (to the point of potentially even being independent) adds a degree of difficulty. Just as with any type of call, VoIP calls are susceptible to static caused by any processing piece of hardware that experiences an electrical short and failing. Echo is also a possible issue and is handled by echo cans (devices that remove echo).

Latency issues

The main problem with VoIP is the negative effects of the delays in transmission of VoIP packets, referred to in the industry as *latency*. A small amount of latency is inherent to converting a call either to or from VoIP. It only takes a few milliseconds to perform that action, but the effects of latency are cumulative over the transmission of the call. A total latency from your Asterisk to the final piece of VoIP hardware at your carrier of less than 200 milliseconds is acceptable. You shouldn't have noticeable effects at this level, and the calls should complete without incident.

Latency of over 200 milliseconds from end to end on the transmission of a call causes portions of the voice transmission to be delayed. The main result of latency is a delay in the transmission. Many years ago, this issue was not uncommon on international calls. As you would speak, a half-second delay would occur before the individual on the far end of your call heard your voice and responded. In the end, she would speak to you while you were speaking (because her voice was delayed about a half-second as well), and you ended up stepping on each other's conversation. This is also an undesirable side effect on some VoIP calls with heavy latency.

Jitter issues

Jitter refers to fluctuations in latency. It is generally the result of failing hardware or a network configuration that must handle not only VoIP but also Internet traffic, e-mails, and office LAN data transmission. Isolating the Asterisk server on its own Internet connection reduces the interference by other applications. You can use the jitter buffers on the Zap channels and the IAX protocol, but the problem shouldn't get to that level.

The result of jitter on your circuit is most commonly an effect called *clipping*. A normal conversation sounds like this:

"Hi, I am just about to leave work and drive home."

A clipped conversation might sound like this:

"Hi, I a_ ju_t abou_ to lea__ wo_k an_ _rive home."

Jitter causes this phenomenon due to the variances in latency that define it. If the *d* sound in the word *drive* has 2000 ms of latency and the *rive* portion of the word has a normal latency of 200 ms, the *d* sound arrives too late and is discarded. The more your latency fluctuates, the more portions of your audio are ejected from the audio stream, and the clipping is worse.

Unless you are integrating VoIP into a huge network at your office that uses the same cabling to provide the VoIP phones as well as Internet access and e-mail, latency and jitter shouldn't be much of an issue. The only way to reduce your latency in normal Asterisk applications where your IP connection is only used for VoIP is to find a better Internet Service Provider whose network is closer to the Internet backbone.

The voice portion of the call is the source of call-quality issues, so the fewer potential points of failure, the cleaner the call. For this reason, we recommend reinviting your VoIP calls to the farthest endpoint possible. Reinviting eliminates your Asterisk server as a potential source of call-quality issues in the RTP stream. If the voice portion of the call never hits the server, it can't affect the echo, static, or latency on the transmission. The overhead portion of the call is still known to the Asterisk server, but the messaging passed is generally restricted to call setup, teardown, and maintenance messages. See Appendix B to find out how to reinvite your calls.

Battling one-way audio

One-way audio is that strange anomaly where you can hear the person calling you, but that person can't hear you (or vice versa). It is uncommon in the world of telephony but it can occur in VoIP transmissions, simply due to the structure of a VoIP call. Whereas regular phone lines are designed for the two-way transmission of voice traffic, each voice (RTP) stream in VoIP is independent. Should one of the RTP streams be lost or misdirected, the conversation only has one active RTP stream, and one-way audio results.

One-way audio is generally the result of an improper negotiation of the RTP stream by a server with a firewall. If the firewall is not accommodating, the inbound RTP stream is denied and one-way audio occurs. This is also not uncommon when the RTP stream is reinvited and both endpoints are protected behind firewalls. Check one of these places if you experience one-way audio. If one end of a VoIP call terminates behind a firewall, the network administrator for that end must set the Network Address Translation (NAT) parameter on their Asterisk switch. This allows the new port behind the firewall to receive the RTP so both ends can receive and hear the audio in the call.

Troubleshooting outbound VoIP calls

Outbound VoIP call problems are worked through in a manner similar to troubleshooting outbound analog calls. First clear your hardware of fault, and then open a ticket with your outbound VoIP carrier.

You can build a dialplan to allow your calls access to all your carrier options, but that also adds the question of exactly which carrier failed the call and how can you prove that to the carrier. For this instance, you must look at VoIP as a solitary element that is isolated from the rest of the Asterisk options. If you add all the possible analog, digital, and IAX possibilities, you add more elements you must rule out as the problem. As long as the call is leaving your hardware correctly and you're receiving the proper responses from the Edge Proxy Server (EPS) of your outbound VoIP carrier, your network is vindicated.

Load Wireshark or tcpdump software on your Asterisk server to execute a packet capture on the IP address that is sending the failed call. This provides the data required to confirm that your VoIP carrier is receiving the required acknowledgment (ACK) and okay (OK) messages. If the transmission has an obvious problem, your SIP capture has the information necessary to quickly resolve it. To find out about packet capture software and understand how to use it, check out Chapter 10.

Assisting with Wireshark

All VoIP issues can be tricky to troubleshoot. One of the best tools available to you is Wireshark (www.wireshark.org). Wireshark allows you to read Session Initiation Protocol (SIP) captures in a logical and helpful manner. Reviewing the information in an SIP capture greatly reduces the troubleshooting time required to resolve VoIP problems. See Chapter 10 to find out how to use Wireshark.

If the SIP capture is fine and had no obvious issues, attempt the same call from your cellphone. If the call completes, call your outbound VoIP provider and open a trouble ticket on the issue.

If the call fails, troubleshoot it as you would an analog phone (see the section "Working through Analog Issues," earlier in this chapter).

If any of the call attempts complete, call your outbound VoIP provider and open a trouble ticket on the issue. Inform the provider that the call completed over another carrier (letting the technician know which one you used). He can place the information into its system and begin to resolve the problem.

Addressing incoming VoIP call problems

Incoming VoIP calls are a different creature than outgoing VoIP calls. Not only do you probably have a different carrier for your inbound VoIP service, but the number that people are dialing into must also be owned by that inbound VoIP carrier.

Phone numbers used by inbound VoIP carriers are generally called Direct Inward Dial (DID) numbers. This is because they can only legitimately be used to receive calls. You can use your Asterisk to brand an outbound call with the DID number as the origination caller ID, but that is more of a programming sleight of hand than a feature inherent to the number. The DID numbers you need to know are as follows:

- ✔ Existing DID numbers
- ✔ New DID numbers
- ✔ Migrated DID numbers

Each variety of DID numbers has its particular challenges and potential pitfalls. Existing numbers are the easiest to troubleshoot, with migrated numbers being the most difficult.

Working with existing DID numbers

Existing DID numbers are active phone numbers on your Asterisk server. The carrier has activated them on your switch, and you've successfully received calls in the past. After a DID number is active on your server, it shouldn't stop working unless you cancel it.

If you receive any type of failure recording or fast busy signal when dialing a DID number on your Asterisk, set up an SIP capture and investigate what is happening. The capture can immediately tell whether your Asterisk is rejecting the call or whether the call isn't even reaching your server.

You can also view the call activity from the Asterisk CLI. You may not be able to see the detail available from a packet capture, but you can identify whether Asterisk has received the call. Servers handling a high volume of calls become a forest of data, making it difficult to find a single call. Access the Asterisk CLI first, and then make a test call into your server. It is much easier to locate the call as it is occurring in the CLI than to hunt for the call after it is complete.

If the call is being rejected by your Asterisk, check your programming for the DID number. If you are using Asterisk in a cluster, the routing for the DID number may simply never have made it into the server. The SIP trace indicates the IP address to which the call was sent, and you can begin your investigation there.

If your SIP capture indicates that the call never hit your Asterisk server, call your inbound VoIP provider and open a trouble ticket. It should be a simple process, where the VoIP provider updates the national routing for that specific DID number.

Small problems can be a big pain. A DID number that is working fine can also experience problems because of an update, or failure to update, the national database. If your phone number moved to an inbound VoIP carrier such as Level 3, the national database for that number must be updated to reflect Level 3 as the new owner of the number. If a local or long-distance carrier fails to gather the new updated information, calls from the location to your DID number fail. The challenge is the fact that your carrier is Level 3 (in this example), when in fact the local carrier is misrouting the call.

Because you don't have a direct business relationship with the offending local carrier, you don't have leverage with it. You can't reference the Service Level Agreement (SLA) in your contract to force the carrier to correct the problem in a timely manner because you don't have a contract. Your inbound VoIP carrier must negotiate with the offending carrier. To make matters worse, the local carrier that is misrouting the calls may use another carrier to reach your inbound VoIP provider. Now a third party, which is in the middle of these two carriers, may have no motivation to fix your problem. Stay persistent, calm, and friendly, and things resolve in the most efficient manner possible.

Adding new DID numbers

You can order new DID numbers from your inbound VoIP carrier. This can cost from 25 cents to $5 per month per phone number; these numbers are doled out to all local carriers in a lottery system once a month.

Before the DID numbers can effectively reach your Asterisk server, the carrier must activate and route them to you. You generally receive notification of the numbers after they have been provisioned to your server and they are ready to go.

When you receive the new DID numbers, follow these steps to get them up and running on your Asterisk server:

1. **Add the DID numbers to your Asterisk dialplan.**

   The calls fail if the server doesn't know what to do with them. Even if the numbers haven't been fully activated with your VoIP carrier, you still want your server to be ready for them.

2. **Dial the DID number to confirm that it completes.**

   If your call receives anything other than an expected answer, set up an SIP capture and dial the DID number again. The information obtained indicates whether the number hit your IP address or whether nothing came through. If the call hit your IP address, investigate the capture to see why the call was rejected. If the call never touched your hardware, call your inbound VoIP carrier to determine whether the number has been fully activated yet.

Unraveling migrated DID numbers

Numbers in the process of migrating from one local carrier to your new VoIP inbound carrier are nervous creatures. One day they have caller ID, have call waiting, and are part of a business group, and the next day they are stripped of all their features and sent off alone to a new local carrier. These numbers exist in a *liminal* (betwixt and between) stage, while paperwork and information are passed between the two local carriers.

If you have problems with a phone number that is either in the process of migrating or has just recently migrated, follow these steps to identify the carrier causing the issue:

1. **Execute an SIP capture on the failing number to check whether the number is hitting your server.**

   If the number is hitting your network, check your dialplan and ensure that it is added everywhere that is should be. If the number isn't hitting your Asterisk server, proceed to Step 2.

2. **Identify who owns the phone number.**

   You want to be chatting with the correct carrier if the number is failing. Call your new VoIP inbound carrier and ask which carrier owns the number that is failing. Because you are moving traffic to this carrier, it has the greatest incentive to help you.

3. **Contact the carrier that is responsible for the phone number.**

   The phone number may have been sent across to the new carrier early and it wasn't ready for it. The other possibility is that the number was canceled when other numbers on your account were migrated. Regardless, you must engage the carrier who owns the number to negotiate the release or repair of the number.

Handling IAX Issues

InterAsterisk eXchange (IAX) transmissions are slightly different than VoIP transmissions with SIP. They are still VoIP calls, but they are a different variety of VoIP, such as H.323 or Media Gateway Control Protocol (MGCP). IAX calls can be reinvited, but the catch is that the entire call is reinvited, not just the RTP stream. Any server in the middle of the call loses track of it when the reinvite is executed. IAX transmissions also use a single port for both the voice stream and the overhead. The likelihood of one-way audio on IAX calls is reduced, but the following issues can crop up with these transmissions:

- ✔ **IP traffic not hitting your machine correctly:** If the IP address or port isn't programmed correctly, the transmission isn't received. You can easily rectify this problem by validating the IP address and port in the configuration for both ends of the call.

- ✔ **Username/password issues:** This type of problem shows up in the Asterisk CLI as a *login failed*. VoIP calls are rejected with a code of either *401 unavailable* or *403 bad auth/forbidden*. Correct the login information, and the call completes.

✔ **Network issues:** The same problems that can affect VoIP and regular IP transmissions can also affect IAX transmissions. Review your router and conduct some testing with your Internet Service Provider to resolve these issues.

✔ **Firewall concerns:** Firewalls aren't as much of a problem as they are with other protocols. The fact that the overhead and voice portions of an IAX call are sent from the same port identifies the issue quickly. Either the call connects and the audio is sent or the call fails. This all-or-nothing aspect of IAX transmissions allows you to easily identify and resolve firewall issues.

Aside from these few mundane problems, IAX is easy to set up and use. The Asterisk CLI is your best tool to identify and resolve issues. It allows you to watch the call progress through all the steps from start to finish. If the problem isn't related to the IP transmission or username/password, it is generally dialplan related.

Debugging Your Devices

You will probably encounter all the telecom problems listed in this chapter, along with issues related directly to your Asterisk. Initially, you will encounter more issues related to hardware devices that are improperly configured or devices for which the wrong drivers are loaded. Debugging the software is straightforward. Execute the debug command from the Asterisk CLI for the hardware device in question and you get a plethora of data. Follow this syntax for a successful debugging session:

```
channeltype debug
```

You can't debug an analog port on a Zaptel card by specifying an IP address, so minor modifications are required for a specific device. For instance, you can issue a general VoIP debug command or one that is specific to an individual IP address, as follows:

```
sip debug
sip debug ip xxx.xxx.xxx.xxx
```

The resulting information illuminates dialplan problems. These issues generally result from a misdirection to an extension or a priority other than where you intended. This can also be seen from the Asterisk CLI on a server without much traffic. After understanding the source of the misdirection, you can repair the coding in your dialplan and correct the issue.

You can use verbosity to determine different levels of information. The higher the number, the more information Asterisk gives you.

A verbose level of 1 provides general information on the device in question, as shown in the following code:

```
Set verbose 1
```

A verbose level of 10 provides in-depth information on a device, allowing you to see the path taken by a call within your Asterisk:

```
Set verbose 10
```

Chapter 13

Handling Dedicated Digital Troubles

The two previous chapters cover everything you need to know for troubleshooting analog, IAX, and VoIP calls. That leaves the digital interface, which we cover in this chapter. The interface card you're using may be from Digium, Sangoma, or Zapata, but all cards perform the same function. They receive a dedicated circuit with 1.544 Mbps of bandwidth and convert it into 24 individual channels that are capable of handling the transmission of phone calls. The interface card may be digital, but in the world of telecom, the line it receives is called a *dedicated circuit*.

Dedicated circuits are unique creatures in telecom. They exist in a realm where more information is available to your carrier, but less may be readily apparent to you. Troubleshooting dedicated circuits can be frustrating, as small nuances of a configuration can result in the complete failure of your circuit. The unique and vital nature of dedicated digital circuits demands its own chapter on troubleshooting.

Identifying the Level of Your Problem

Effective troubleshooting requires a plan. Without a methodical and cogent way of identifying, isolating, and proving out all the variables on the circuit, you can easily spend weeks chasing your tail and accomplishing nothing.

Dedicated circuits can come in many sizes, but the digital interfaces for your Asterisk are only available in two increments: DS-3 and DS-1. A DS-1 can handle 24 individual calls (on channels called *DS-0s*) and is the basis for every dedicated circuit, regardless of size. A DS-3 is a larger circuit than a DS-1 and contains 28 DS-1s.

The terms *DS-0, DS-1,* and *DS-3* are international terms that refer to the levels of circuits. The DS-1 represents the smallest complete digital circuit available. In North America, this circuit is called a T-1, which includes the 24 DS-0s. In Europe, an E-1 contains 32 individual DS-0s. Generically, they are DS-1s. The same goes for DS-3; it includes 28 DS-1s, regardless of whether they are E-1s or T-1s.

Identifying the level of a problem is the first step in the troubleshooting process. DS-3–level issues have a different set of variables than DS-1–level variables. Only carriers and hardware that interact at the DS-3 level should be addressed in handling DS-3–level problems. Here's a rundown of the basic levels of circuitry you might be dealing with:

- **DS-3:** Problems at this level affect all the lower-level DS-1s and DS-0s. If red lights or alarms are flashing on your DS-3 multiplexer where the coax cable enters, and all of your DS-1s are down, you can trace the source to the DS-3 level.

- **DS-1:** Problems at this level affect the entire DS-1 circuit and all the individual calls running on the circuit. If you have an individual DS-0 channel that is unaffected, your problem is not at the DS-1 level.

- **DS-0:** Problems at this level affect the individual call channel. These problems are either isolated to individual DS-0–level interfaces on your Asterisk or the DS-0–level hardware at your carrier, or they are somewhere within the Public Switched Telephone Network (PSTN — see Chapter 11). You know you're dealing with a DS-0 issue if the problem only affects specific calls and not every channel on the entire DS-1 circuit.

The good news is that troubleshooting anything above the level of a DS-1 circuit involves the same carriers and quantity of hardware. The greatest change in variables occurs when troubleshooting issues at the DS-1 level and individual-channel (DS-0) level.

If your local carrier provides your dedicated circuit, be sure to include the carrier in your troubleshooting process. Most dedicated circuits for voice calls terminate at a long-distance carrier, and because of that, the local carrier has only a supporting role in the circuit. As you isolate dedicated circuits from the DS-3 or DS-1 level down to the individual DS-0 channel level, take note of the variables at work and who is responsible for each one.

Beware of the misdiagnosed problem

Begin troubleshooting your dedicated circuit only after you confirm that your issue doesn't exist anywhere else. Compare call types and confirm the source of your telecom problem (see Chapters 11 and 12 if you need help doing so). Your dedicated circuit only involves a few miles of cabling and a few pieces of hardware between your company and your carrier. In spite of the fact that each channel is broken down by the Zaptel card, be sure to differentiate problems with the T-1 circuit versus issues within the switched network beyond. In fact, you may make your telecom universe more painful if you open a trouble ticket for your dedicated circuit

when in fact the problem could be effectively resolved as an analog issue. Put simply, troubleshooting the dedicated circuit places it at unnecessary risk; you may end up taking down the circuit, leaving your business without phone service for the duration of testing.

If your issue affects a variety of your long-distance calls, from switched outbound calls to dedicated inbound calls, we advise you (nay, we *entreat* you) to open your trouble ticket on the switched outbound issue because it provides the greatest focus for your carrier.

A dedicated circuit is only dedicated from where it connects into your Zaptel T-1 card to the point that it terminates into your carrier's muxing hardware that splits it into 24 channels. After the circuit is split into individual channels within your carrier, the calls are processed and routed just as if they had arrived from a regular analog line. You should handle any problem beyond that point as if it was an analog problem.

Differentiating local from long distance

You can identify dedicated digital circuits by the carrier into which they terminate. If you order a DS-1 circuit from your local carrier, it is referred to as a *local circuit*. Similarly, a DS-1 circuit from your long-distance carrier is referred to as a *long-distance circuit*. You must realize the difference, because long-distance circuits don't have the same functionality as local circuits.

You need a local circuit for the following reasons:

✔ **To process outbound toll-free calls:** Long-distance circuits can't process outbound calls to toll-free numbers. Long-distance switches (called *Class 4 switches* in the telecom world) into which long-distance circuits terminate don't have this functionality. The Class 4 switches don't have the required infrastructure to identify the owner of the toll-free number and route it to that carrier. Long-distance dedicated circuits can receive toll-free calls; they simply can't dial out to toll-free numbers.

- ✓ **For 911, 411, and 611 services:** Long-distance circuits aren't set up for these services. You can only get these features available through the local carrier switch (called a *Class 5 switch*).

- ✓ **To dial a 1010 dial-around code:** Just as your long-distance carrier can't accept toll-free number calls and send them to the carrier that owns the toll-free number, it also can't route your call to another long-distance network if you attempt a dial-around code. Any attempt to do so ends up in a fast busy signal or, as Asterisk sees it, *congestion*.

You need a long-distance circuit for these reasons:

- ✓ **To receive calls on a standard phone number:** Your Zaptel T-1 card can receive an inbound call on a dedicated long-distance circuit only if a toll-free number is routed to it by your long-distance carrier. (Again, we are talking about inbound calls to your circuit, not outbound calls from it.) Local T-1s have more flexibility in that regard and can have inbound-only phone numbers called Direct Inward Dial (DID) numbers attributed to them. This allows anyone to call a seemingly normal phone number to reach your Zaptel T-1 card.

- ✓ **To receive dedicated long-distance rates:** Long-distance service has two primary tiers of rates: analog (or *switched*) and digital (or *dedicated*). The dedicated per-minute rates are frequently half the switched rates, but they are only available to you on a dedicated long-distance circuit. In spite of the fact that the local digital circuit is dedicated, it looks just like an analog phone call by the time it hits your long-distance carrier. These calls aren't eligible for the reduced dedicated rates. This is why people order long-distance circuits that can't call toll-free numbers or 411.

A problem must involve all your T-1 circuits for it to be a true dedicated issue. If you have echo, static, or a failure to complete when dialing to only a single phone number and calls to all other phone numbers are fine, report it to your carrier as an individual call issue.

Problems that are truly associated with your dedicated circuit include the following:

- ✓ **Static on every channel after you dial a phone number:** The dial tone you hear from your Asterisk when you pick up the phone is provided by Asterisk. Only after the last digit is dialed is the call forwarded through your dedicated circuit. If you hear static or echo before you even begin dialing, the problem is most likely a failing piece of hardware in your Asterisk server, not a problem with your dedicated circuit.

- ✓ **Circuit down hard:** This disposition is indicated in the Asterisk command-line interface (CLI) in the Linux operating system and as red lights illuminated on the back of the Zaptel T-1 card. If the circuit is simply impaired, but not down, yellow or blue alarm lights are lit.

Asterisk doesn't care about alarms on the T-1 cards. The channel is understood by Asterisk to be active (green), down hard (red), or impaired (yellow or blue). No further information on the disposition of the circuit is inherently known to the Asterisk dialplan.

Additional information is available on your dedicated digital circuit by executing the following command from the Asterisk CLI (replace the *X* with the number of the port or span):

```
zap debug pri span X
```

This command identifies the configuration and disposition of the dedicated circuit. If your programming appears fine and you don't suspect issues with the cabling between your Asterisk and the wall jack, call your carrier and report the trouble.

Identifying DS-1–level circuit variables

A problem on a single channel can be associated with the dedicated portion of the circuit. If the echo, static, or call failures are associated with only one port on a T-1 card, the problem can only exist in one of the following three places:

- ✔ **Your Zaptel hardware:** Every card and piece of hardware can fail — even a single port on a card. Some people claim that a T-1 card fails if an individual channel on it goes bad. But these same cards can pass calls on 22 channels without incident and only fail on two channels. Take all claims of the all-or-none reports from hardware manufacturers with a grain of salt.

- ✔ **The multiplexing card at your carrier:** Dedicated circuits are like math equations: Whatever happens on one side of the equal sign also has to happen on the other side. A card in your carrier's switch breaks the circuit into 24 usable channels in your server — which is also a function of your Zaptel T-1 card. The card may be called an SPM, SPMI, or DTC card, but they all perform the same function. A problem with an individual channel can only be the result of the failure of a piece of hardware that interacts with your circuit at that level.

- ✔ **An echo canceller in the circuit between these two points:** Carriers know that you don't like to have a phone conversation that sounds like you're in a marble bathroom. To prevent the echo that's possible in the transmission of calls, the carriers install devices called *echo cancellers* (or *echo cans*) in the circuit. These devices interact at the individual channel level and can fail, causing static, call failures, and even echo.

A hardware issue in one of these three places can cause problems with the individual channel. The echo cans are the easiest to troubleshoot because they can usually be temporarily removed from your circuit by your carrier. If the problem persists on the channel, you know that the echo cans aren't the problem.

T-1 issues are generally worked by carriers through a process of looping. *Looping* refers to the process whereby an electrical signal is sent down the circuit to test for continuity and quality. The downside of looping is that the circuit is unable to send or receive calls while it is being looped. Because the circuit is unavailable during these loop tests, the process is called *intrusive testing*.

Nonintrusive testing doesn't exist. Maintenance files on circuits collect errors, but they don't indicate the location of a problem. They simply validate that an undesirable condition occurred. They don't tell you when the errors were generated, but simply that they were collected between the moment they were read and the last time the error file was cleared. That could have been minutes, hours, or days.

Identifying DS-0 or individual channel issues

The main variables at the DS-0 level are the multiplexers on either side of the circuit that break the T-1 into individual channels. If your third DS-0 on your T-1 has static or is unavailable, or if you can't get a dial tone on it, either the Zapata DS-1 card in your server or your carrier's hardware is failing for that channel. If you can isolate both pieces of hardware and the problem doesn't clear, the only other likely candidate is an echo canceller somewhere in your circuit that is malfunctioning or slowly failing. If the device is failing or was provisioned incorrectly, it can prevent your call from completing with good quality (or completing at all).

Files saved by computer software to monitor call performance are unoriginally called *performance monitors (PMs),* and they collect information on the quantity and type of errors on a dedicated circuit. Request the information from your carrier by calling to open a trouble ticket. PM software groups errors in several categories. If no error was found on a circuit in a particular category, a 0 error count is listed. The quantity of each type of error guides you to other areas of inquiry when you see them. Here are some common PM errors:

✔ **Erred seconds:** You find *erred seconds* on a circuit with a minor problem. An erred second indicates that for the duration of one second, the communication of the circuit was distorted. The overhead information may have been lost for that specific second because some piece of hardware experienced a low-level, intermittent electrical short. You might experience minor static or quality issues that are more of an annoyance than a concern.

✔ **Severely erred seconds:** You can find severely erred seconds on a circuit with large issues. The severely erred seconds indicate a larger issue whereby information was not only missing but also replaced with aberrant data. These errors often cause static or line noise so severe that you can't hear the person you are calling. If your circuit has severely erred seconds but you aren't experiencing regular quality issues, you might instead notice that your calls disconnect prematurely or that your circuit drops unexpectedly.

✔ **Framing slips:** Framing slips signal a configuration problem, generally a timing issue where your hardware is attempting to correct a lag behind the master clock of your carrier. If you aren't *clocking* off your carrier (using its timing as the point of reference on the circuit) or if your hardware is set up for a configuration that doesn't match your carrier, you can expect frame slips.

The bad news about frame slips is that they may be minor, and go unnoticed by you as they accumulate, up to the point where they drop your circuit.

✔ **Unavailable seconds:** Unavailable seconds are just that: seconds of time when the network thinks your circuit is unavailable. Unplugging your DS-1 circuit from your Asterisk generates unavailable seconds in your carrier's network as easily as a fiber cut or an accidental removal of your cross-connect by your local loop provider.

One of the simplest things you can do to see whether a problem on your dedicated circuit can resolve itself is to reboot your Asterisk server. If you decide to reboot, make sure that your carrier pulls the PMs before you reboot your hardware; this way, you can prevent confusion. Even if you have a clean circuit, as soon as you power down your server and reboot it, you create a handful of errors on the circuit. Despite the fact that you have a legitimate issue on your circuit, rebooting has the potential of masking the true problem, enlarging the number of errors your carrier finds.

Without controlled testing, your carrier can't know which piece of hardware is generating the errors or when the errors were created. Errors that exist in the PM files don't have time stamps next to them. Only total error counts for a given period of time are listed. If the PMs show that your circuit received 176 erred seconds and 56 severely erred seconds in the past 24 hours, you have no way of identifying whether all the errors occurred in one group as your circuit took a hit or whether they were generated periodically over the past 24 hours. The only way to validate the frequency of errors is to purge the PM file and check it every 15 minutes.

Opening a Trouble Ticket for Your Dedicated Circuit

After you have some preliminary information about the trouble with your circuit, call your carrier and open a trouble ticket. Your carrier needs to know some reference information from you to open the trouble ticket, so ensure that you have your account number and circuit ID ready before you make that call. In the perfect world, either you or your carrier would provide you with a technical cut sheet on your circuits for just such an occasion. This reference becomes more helpful as you collect additional circuits with potentially different configurations. A fully fleshed-out cut sheet includes the following items:

- The circuit ID assigned by your carrier
- The trunk group name assigned by your circuit
- The order number attributed to the circuit by your carrier
- The account number attributed to circuit
- The complete configuration of your circuit
- The trouble-reporting number for your carrier
- An escalation list for trouble-reporting issues with your carrier in case you can't reach the main office or it fails to meet the established response times.

Going through the basics

If you don't have a cut sheet, at a minimum, you need to write down the circuit ID for every circuit that is having problems. Your carrier needs this information to identify your circuit and begin the troubleshooting process. In addition to your circuit ID, your carrier might also ask you the following questions:

✔ **Does your Asterisk show any alarms?** If so, what color are they? Both the alarms and their locations indicate to your carrier the kind of problems your circuit is experiencing. If you have red alarms on your Asterisk, the carrier knows that the circuit is down hard and that your business has no phone service. If your Asterisk is experiencing yellow alarms, your circuit could be intermittently down and bouncing back. If you don't know this information at the time you call into your carrier, you can simply tell your carrier that you haven't checked yet, but that you still want to open the trouble ticket.

✔ **Have you rebooted your Asterisk server, and if so, how did it affect the problem?** You might temporarily resolve some problems by rebooting your server. Rebooting enables the digital interface card to refresh the connection to your carrier and sync up. The only variable you are testing when you reboot is the interface card. If the problem clears when you reboot, your hardware is probably the source of your issue. If it persists without change when you reboot, the source of the problem could be within your carrier.

✔ **What are your hours of operation?** In the event that a technician has to be dispatched to your office to fix the problem, your carrier needs to know when someone can be there for the technician. If you have an after-hours employee who will be on-site, be sure to provide that employee's name and phone number (if the system is down hard, provide a cellphone number).

✔ **When did you first notice the problem?** This information gives your carrier direction in the search. If a large outage hit your area an hour ago, and that is roughly when you noticed the situation, the carrier can easily combine your issue with the overarching trouble ticket (sometimes called a *master trouble ticket*). Of course, your problem may not be related to a larger issue, but offering a time frame gives your carrier some indication of events that occurred that may have caused your issue.

✔ **Who is the site contact?** Your carrier needs to know who to call for updates. Provide a direct phone number, a cellphone number, and a secondary means of contact if your contact plans on being unavailable at any time during the day.

✔ **Is the circuit released for intrusive testing?** *Intrusive testing* is the most direct way your carrier can investigate an issue with a dedicated circuit. The main downside of intrusive testing is that your circuit is down while your carrier takes over every channel. Of course, if the system is already down, you have nothing to lose. On the other hand, if your circuit is active when you release it for intrusive testing, active calls are disconnected when the carrier initiates the test.

Letting your channels be your guide

When you open a trouble ticket, it's always helpful to ask the carrier for a *circuit snapshot*. A snapshot of the circuit is the disposition of each channel on your circuit. Even if you think you have a DS-1–level issue, it's always a good practice to ask for this information. You might believe that your entire circuit is being affected, but a snapshot could reveal that only half the channels are impaired. With basic facts from the PMs and a circuit snapshot, you can begin setting realistic expectations for resolution of your service. Additionally, you can isolate individual DS-0 issues to the specific channels that are experiencing the problem. To solve your problem, you might be able to reboot your server to bring the channels back into service or modify your dialplan to deselect them as available lines. The key is using every available bit of information your circuit provides. Taking in the disposition of your individual DS-0 channels on your DS-1 circuit is a quick and easy way to glean what is happening to your circuit.

Idle: IDL

Your channels are in an *idle (IDL)* state when they have access to your carrier's network but don't have an active call. This is the state you expect your channels to be in when everything is working fine and the circuit is waiting to process a call.

The circuit must be considered idle by both your carrier's technology and your own phone system for a call to come through on the line. If, according to your carrier, the channels are idle, but your hardware is unable to seize the channels for some reason, one end of the circuit isn't speaking to the other. Open a trouble ticket. A challenge might exist between your Asterisk software and the Zapata digital interface card. Regardless of the source of the issue, opening a trouble ticket with your carrier and using it to test your cards are the most efficient way to resolve the problem.

Call processing busy: CPB

Call processing busy, or *CPB,* is the healthy state of a channel with an active call on it. Under normal circumstances, your active circuit has several channels in the CPB state, indicating active calls on the specific DS-0s. The rest of the channels are idle, waiting to accept a call. A circuit with channels in CPB is obviously connected to the carrier because only a call that is connected to the network can establish the channel in this state.

Remote made busy: RMB

Remote made busy, or *RMB,* is a common disposition for a channel if you're dealing with hardware issues. The carrier may see a DS-0 in this state when your interface card locks out the channel. Channels in an RMB state can sometimes be corrected if your carrier takes down the specific channel and resets it (this process is commonly called *bouncing a channel*). If your carrier can't bring the DS-0 back into service by bouncing it, the only way to restore the channel is for you to reboot your server. If this condition is chronic, contact the manufacturer of your digital interface card for guidance with testing or purchasing a replacement. Finding the problem when it's minor is better than letting it go to the point at which the channel won't come back to life after you reboot.

D channel made busy: DMB

The *D channel made busy (DMB)* state is used only with ISDN circuits. Only ISDN circuits have D channels designated for signaling, and that is why non-ISDN circuits can't have channels in this state. When the D channel made busy state appears on an ISDN circuit, the entire circuit is down. The D channel handles all the logistics of call setup and teardown on an ISDN circuit, and, without the overhead, the circuit can't function.

This condition can be the result of your digital interface card having issues similar to an RMB condition. Other possible sources of DMB are your carrier having issues and an ISDN protocol mismatch.

Installation made busy: IMB

Installation made busy, or *IMB,* is a protected state that is usually imposed by your carrier on an entire circuit to prevent the circuit from generating alarms in its network. For example, if you unplug your Asterisk server from the circuit, your carrier immediately sees alarms at its end of the span. The technician who is listening to the piercing screech of the alarm doesn't have your contact name and phone number, so he or she does the easiest thing to end the maddening alarm. Usually, the easiest thing to do is to *busy-out the circuit,* or place the circuit into the IMB state. The bad news is that when you plug your hardware back in, you can't use it because the carrier has busied it out.

Every carrier has its own procedure for placing a circuit into the IMB state. The process might be manual, with a policy to wait until the circuit has been in alarm mode for eight or ten hours before busying-out the circuit. Sometimes, however, the process is automated; you may have as little as an hour before the carrier automatically busies-out your system.

Avoiding permanent IMB status

To prevent your circuit from spending more time in IMB than it needs to, open an *information-only trouble ticket* with your carrier before unplugging your hardware for any reason. This type of trouble ticket alerts the carrier to the fact that you are removing your hardware from the circuit and gives you a trouble ticket number to reference when you call to have the circuit taken out of IMB.

Removing a circuit from IMB takes about five minutes after you reach a technician with the required credentials to make the change. The greatest potential delay comes from wading through the customer service department and waiting for a certain technician with access to the switch to call you back.

Carrier failure: CFL

A *carrier failure (CFL)* state is like IMB, because it generally affects your entire circuit, not just a few channels. The bad news about a circuit in carrier failure is that this state commonly indicates a substantial problem that needs to be addressed immediately by your carrier. If your CSU fails, if someone accidentally cuts through the wiring of your circuit with a backhoe, or if your carrier's switch sustains a direct lightning strike, your circuit is in carrier failure.

Carrier failure indicates a more serious issue and generally takes several hours to diagnose. After the technician finds the source of the problem, it can be another 4 to 24 hours before your circuit is back in service. Keep this in mind when you set expectations. (In other words, it's time to have a chat with the head of your telemarketing department.)

Using a T-1 test set

If all the previous tests don't reveal the source of your problem, your next step is to replace your hardware and see whether that brings the circuit up. If this doesn't resolve your issue, you need to call out a hardware technician with a portable *T-1 test set,* possibly a Phoenix test set Model 5575 or a T-Berd test set made by TTC/Acterna (who was just purchased by a company called JDSU out of Milpitas, CA; www.jdsu.com).

If your business is providing phone service with the Asterisk software functioning as your backbone, we recommend buying a solid piece of testing hardware. Test sets look like a cross between a toolbox and a 1930s lunch box. They range in price from a few hundred dollars for a used Phoenix 5575A to several thousand dollars for a fully fleshed-out TTC T-Berd or TTC FireBerd with all the available ISDN options.

Having a test set for your dedicated circuit is helpful, so chat with your hardware provider to determine the best test set for your application and budget. A refurbished Phoenix 5575A might be all you need and is well worth the $200 you might spend for it on eBay.

Managing Your Dedicated Trouble Ticket

After you open your trouble ticket with your carrier, be sure to write down all the following information to make it easier to manage the issue:

✔ The trouble ticket number.

✔ The name and contact information of the person at your carrier who opened the ticket.

✔ The time and date the ticket was opened.

✔ The disposition of the DS-0s of your circuit (if available).

✔ The errors listed on the performance monitors, if any (if the customer service representative can access the file).

✔ Any testing you completed and the results of the tests. (Did your carrier bounce the circuit by taking the circuit out of service and restoring it? Did that solve the problem?)

The final bit of information your carrier should provide you is when you can expect a callback on your ticket, and who will be calling. Regardless of the time frame you're given, write it down and schedule a callback to the carrier. If the technician calls you before the deadline, that's great, but if he or she doesn't call, you need to call back to escalate your issue.

Reducing the Impact of an Outage

Asterisk has the good fortune of being able to access Internet connections for VoIP calls as well as analog channels. This provides alternate ways to route your calls if one avenue (in this instance, your digital circuit) is unavailable.

Create a dialplan with fallback capacity in the event that a resource, such as an analog or digital Zaptel card, is unavailable. The ChanIsAvail() application confirms whether a specified channel is available before sending a call to that extension. In a dialplan, the code appears as follows:

```
exten=>2565551212,1,ChanIsAvail(Zap/g0)
exten=>2565551212,2,Dial(Zap/g0)
exten=>2565551212,102,Dial(SIP/Saturn)
```

This code confirms whether the Zap/g0 resource is available before sending a call to it. If the resource is available, the second priority dials the Zap/g0 device. Priority 102 dials the SIP/Saturn device for the call if the `ChanIsAvail()` application is false. If Priority 1 is false, the call forwards to priority 102 (representing the existing priority plus 101).

With this type of fallback routing, you can't easily identify which carrier received the call. If you are dialing a phone number that is usually sent out over your analog lines, but the dialplan has backup routes built in, which carrier did it complete or fail over? If your analog line was busy or down, the call may have been routed over a VoIP line or a dedicated digital line.

To open a trouble ticket and resolve an issue, first identify the carrier that is potentially failing the call. Fortunately, you can view a call sent to a dedicated circuit from the Asterisk CLI. Executing the following command from the Linux console connects you to the Asterisk CLI:

```
asterisk -r
```

The command shows the active channels on your Asterisk and the specific end devices that each call is using (as shown in Figure 13-1). In this figure, the call is sent to the VoIP channel named SIP/209.247.16.211-b1600018.

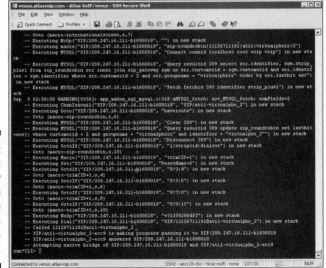

Figure 13-1:
You can identify any messages coming to the console from the Asterisk CLI.

If you're not sure how to maneuver around in Linux, check out Appendix C for all the basics.

Chapter 14

Managing Asterisk for Peak Capacity

*U*nless you are a VoIP-based carrier using Asterisk as the basis for your ever-expanding telecom empire, you won't need detailed tracking and analysis of your call volume. The bottleneck you are encountering is more typically related to a limitation of your server, rather than a limitation of your VoIP, digital, or analog connectivity.

Delegating individual servers to specific applications spreads the load of processing calls. The makeup of your business dictates your Asterisk configuration. If you're using Asterisk equally as a phone system and conference-calling platform, investigate setting up one server for each application.

A call potentially has two prime points of failure due to rampant growth of your company: capacity limitations between you and your carrier, and internal bottlenecks within your LAN. We cover these failures in this chapter as well as describe how to avoid them in the first place.

Handling Carrier Connectivity Bottlenecks

Depending on the connection you use, you have different limitations to work around to keep your Asterisk running at its peak capacity.

VoIP limitations

VoIP connections are much more dynamic than standard analog or digital connections. The quantity of calls available on a VoIP connection is limited by the following:

- Your network capacity
- The codec(s) you are using
- Your VoIP carrier's total line capacity

These three variables allow more flexibility in the quantity of calls possible between your Asterisk and your VoIP carrier. Even if you only have 1.544 Mbps of bandwidth (a T-1 line) to your Internet Service Provider, you may still be able to pass over 50 active calls if you use a compressed codec such as G.729 and your carrier allows you 50 ports.

You may have sufficient bandwidth to process a VoIP call, but if you run out of G.729 licenses, this causes some problems. G.729 VoIP calls are first renegotiated to a different codec (such as G.711) to maintain the integrity of the call. If the remote server doesn't accept anything but G.729, which is uncommon, the call is connected and lacks audio transmission. The codec only processes the Real-Time Transport Protocol (RTP) stream of the call and not the overhead. The resulting call completes because the overhead is working fine, but the codec can't convert the audio, so it fails.

To manage your G.729 usage, you can add information to your dialplan as well as distill your Call Detail Record (CDR) information with MySQL (see the nearby sidebar).

If you find you have insufficient bandwidth between yourself and your carrier, you have only one solution: buy more. Depending on the type of IP bandwidth you've ordered, you may be able to increase capacity in a few days. This is a much shorter timeline than the 30 to 45 days necessary to bring in another digital T-1 line.

Analog and digital limitations

Analog and digital connections don't have as many variables as VoIP connections. Only the installed lines are available to you. If your local carrier installed four analog lines, you can only receive four calls, regardless of how new-fangled your Asterisk server may be.

Managing your G.729 codec with MySQL

By appending a tag on G.729 calls in the UserField section of the CDR, you can differentiate these calls from every other VoIP call. Perform a few queries in SQL, and you have a breakdown of calls per hour or minute that used the specific codec.

The queries required to accomplish this task are based upon the destination channel for each call

that is also parsed out by hourly timeframes. (Check out Chapter 7 for more information on the fields located in the CDR.) This coding required to execute the queries is very complex. We suggest you outsource this task to a programmer.

Most carriers can provide a *lost call report*. This report shows the number of calls directed to your dedicated circuit that failed because all your channels were busy. The report only covers inbound calls to dedicated circuits, so if you request the same information on your analog lines, the data may not be available.

The following dialplan shows how you can track failed outbound calls within your CDR:

```
[outbound]
exten => _NXXNXXXXXX, 1, Dial(Zap/g0/${EXTEN})
exten => _NXXNXXXXXX, 2,
            GotoIf($["${DIALSTATUS}"="BUSY"]?3:4)
exten => _NXXNXXXXXX, 3, Busy
exten => _NXXNXXXXXX, 4,
            GotoIf($["${DIALSTATUS}"="CHANUNAVAIL"]?5:7)
exten => _NXXNXXXXXX, 5, AppendCDRUserField(unavail\;)
exten => _NXXNXXXXXX, 6, Goto(8)
exten => _NXXNXXXXXX, 7, Congestion
exten => _NXXNXXXXXX, 8, Dial(SIP/${SIPTRUNK}/${EXTEN})
```

Priority 5 tags the call with the name unavail in the UserField section of the CDR for the specific call. We used unavail because it works well for us, but you can use any name you want. The more dissimilar the word, the easier it is to find it in a query.

The semicolon (;) is generally used to specify a comment within the Asterisk dialplan. Preceding it with a backslash (\) makes Asterisk read it as a plain semicolon to be added to the data, instead of being a character that specifies the start of a comment. The semicolon after the `unavail` allows you to append multiple comments to the UserField of the CDR and still have a way to easily identify when one comment ends and the next one begins.

We recommend hiring someone to build queries on your CDR in SQL or another database program.

Dealing with Internal Bottlenecks

Internal Asterisk bottlenecks generally occur because you're outstripping the processing capacity of your sever. Handling more active calls requires more memory and horsepower to prevent a slowdown in processor response.

Use the `top` command in Linux to view the utilization of the server. This command displays information on both the processor memory of the server and the processes currently running. If your server is consistently over 50 percent processor usage, or is using a similarly high amount of memory on a consistent basis, it is time to upgrade your server. If you are running the fastest processor available with 4GB of memory and you are still pushing the upper register of your server's capacity, it's time to upgrade to a multiple-server configuration.

Jumping to a multiple-server configuration adds another layer of complexity to the concern about delay. Because Asterisk servers function independently, you need to build your own database software to link them to prevent inconsistencies in call processing. Your unique business application for Asterisk dictates how you manage your servers.

The following two basic types of configuration are possible for a multiple-server configuration:

 ✔ **A main database server and subordinate Asterisk servers:** A *parent-child* scenario establishes one server as the primary database server that receives all modifications to the dialplan or database. The main server contacts all the Asterisk servers in the network, updating their information so that every server is updated efficiently. Figure 14-1 shows how the Asterisk servers are connected to both a proxy server and a database server, but not directly connected to each other. The proxy server receives all incoming calls — VoIP, in this case — and doles them out to the individual Session Initiation Protocol (SIP) servers; the database server updates the Asterisk servers with the latest dialplan information.

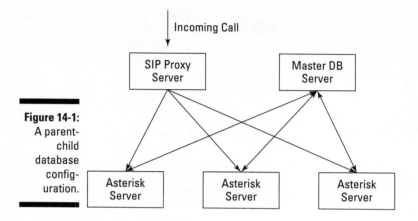

Figure 14-1:
A parent-child database config-uration.

🖌 **A cluster of Asterisk servers:** A cluster of Asterisk servers doesn't have a leader or main server from which all updates are received. Clustering your Asterisk servers allows information to be added to any one of the servers; the information cascades through all the servers. This scenario differs from a parent-child scenario because the Asterisk servers have direct LAN connections between them instead of being separated by either the proxy server or the main database server. Figure 14-2 shows how a cluster of servers works. The inbound VoIP carrier (in this scenario) is responsi-ble for sending inbound calls to each server in turn, instead of the proxy server handling that task as in a parent-child configuration.

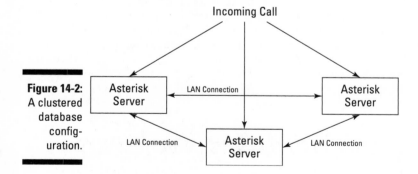

Figure 14-2:
A clustered database config-uration.

Take some time to plan for the future because the option you choose today has far-reaching ramifications for your business. Imagine that you grow 15, 50, or 500 percent in the next few years. You may outstrip the capacity of your main server and have to rebuild the configuration as a cluster. If you're using Asterisk solely for a phone system for your office, you may not want to invest the time and money to build a custom database to cluster two or three servers together that won't ever need expansion. Regardless of your decision, mulling over your possible future growth now is better before you have to reconfigure the entire system than be forced to retool the entire system because you didn't expect so much growth.

Saving Capacity for a Rainy Day

In the perfect world, you would only buy the exact capacity you need. The reality is that some days you need a little more capacity and some days you have plenty in reserve. As a general rule, we recommend having 20 percent reserve capacity available to you at all times. If you're running Asterisk strictly as an internal phone system, you may need to add one or two analog lines. If you're running multiple Asterisk servers that provide VoIP service to hundreds or thousands of customers, 20 percent represents a bit more.

Attempt to complete your calls over your most cost-effective carrier first and then fall back to a secondary or tertiary carrier; this allows you to maximize all your capacity with every carrier.

The per-minute rates charged to you for terminating calls over VoIP may be more competitive than your analog to digital connections, so check out the prices before you select your primary carrier.

This expedient is great for outbound calls, but you can't replicate it on inbound calls. You can't receive a call on your local analog line and have it roll to your VoIP connection with another carrier. The most dynamic solution for inbound is to use VoIP for your inbound calls. It is much easier to open additional capacity on an IP line than to have additional analog or digital lines installed. The benefit of VoIP is that compression allows you to receive and send more calls over the same bandwidth that you use for an analog or digital call.

Entertaining the Third-Party Software

The open-source architecture for Asterisk allows Asterisk to be whatever you want it to be. Third-party management software is great for what it was designed for, but it could always do more. Most third-party software is geared toward single servers running Asterisk and not systems running multiple clusters. Getting the most out of your Asterisk software requires you to either spend the time to become a good programmer or have the money to hire one.

Two of the popular after-market Asterisk software packages are FreePBX and Trixbox (formerly known as Asterisk@Home):

- **FreePBX** (`www.freepbx.org`) is an easy-to use Web-based management portal for Asterisk. Because it's Web based, it's a breeze to set up and prevents you from individually modifying configuration files. As long as you only need the features and setups of what FreePBX provides, it's a great tool. Adding custom features are where FreePBX shows its Achilles heel. The Asterisk configuration files are modified by FreePBX to fit the requirements of the portal software, making it more complex to integrate specialized changes.

- **Trixbox** (`www.trixbox.org`) has similar challenges with retooling the configuration files for custom features. Features it can provide for you depend upon the size and speed of your server. Trixbox contains a group of software programs. These programs are capable of supporting an entire company from database/Web server and Web mail server functionality to contact management and other supporting software.

If you are interested in either of these packages, we recommend that you research their features and choose one that's a fit.

Chapter 15

Providing Long-Term Health for Your Asterisk Switch

Your Asterisk server is a piece of computer hardware just like any other server. As such, you should take care and use precautions to ensure its long-term health and well-being. Some issues — such as hurricanes, tornadoes, and other acts of God — are inescapable, but you can plan for anything else.

This chapter covers the basic environmental concerns to consider for your Asterisk server and provides some software maintenance advice. You probably never intend to reinstall your Asterisk software, but you may have to do that some day.

Using an Uninterruptible Power Supply

Battery backup is essential to ensure the long-term health of your hardware. Small periods of reduced power and the subsequent corrections to full power cause voltage peaks that slowly wear out the silicon in the chips and cards of your server.

Brownouts are periods of reduced electrical flow and are frequent, especially in old buildings. Your lights may dim when you turn on a portable heater or air conditioner. As much as brownouts are irritating, they aren't destructive. The surge of electricity after a brownout ends is what causes the damage to your hardware.

Don't confuse brownouts with *blackouts,* which are a complete loss of electricity.

An uninterruptible power supply (UPS) is a device that can correct a deficiency in power. A drop in electricity through a power line serviced by a UPS is corrected by either accessing batteries to increase the available power or through the use of a voltage stabilizer. A *power conditioner* is a device that removes excesses of power coming through the line, which are generally referred to as *surges* or *spikes.*

We recommend that you always have a power conditioner within your power management system. The power conditioner can either be within the UPS or be a stand-alone device between the UPS and your server. A system without a UPS faces the loss of software and unsaved data during a blackout, and a system without a power conditioner faces catastrophic hardware failure in the event of an electrical surge or spike. Preparing for both possibilities is the safest bet for any device on which your company relies.

UPS systems come in two basic varieties: the standby UPS and the online UPS. Each battery backup system has its pros and cons.

Standby UPSs

Standby UPSs are more economical but aren't as robust as the online systems. Standby UPSs are comprised of an output network that monitors your flow of electricity and a battery backup that's waiting to jump into service. They work by monitoring the flow of electricity for changes. If your voltage drops below 90, the system accesses the batteries and you hear an annoying beeping, but your computer doesn't take a hit.

Standby UPSs have the following downsides:

- ✔ **You may have an extended period where the incoming voltage is less than 90 volts.** The good news is that your battery backup compensates and provides 110 volts to keep everything running nicely. The bad news is that after 25 minutes, your battery can die, and your system returns to the brownout condition. This limitation isn't bad during a brownout, but when you experience a blackout, your system crashes.

- ✔ **The surge of electricity hitting your computer from the UPS cutting over to the battery backup can cause damage to your server.** Just as most light bulbs burn out when you turn them on, and not when they are already on, the surge of power can take a system with minor electrical issues and make it fail. Some lower-quality UPSs have been known to short out the server they were installed to protect in their zeal to maintain electrical continuity.

Creating an agreeable environment

Servers enjoy a cool, dry environment. They work happily below 73 degrees Fahrenheit and below 30 percent humidity. This is easily accomplished by a solid air conditioning system. Keep the air conditioning maintained and change the air filters often; this maintenance can make a substantial difference in the system's ability to cool your server room. Aside from maintaining the temperature, the next biggest concern is maintaining the quality of your electricity.

Online UPSs

Online UPSs are the top of the line. They're generally the most trustworthy — and the most costly.

They not only monitor the electricity that flows through your system, but they also interact with it. Many online systems have built-in voltage stabilizers that bump up the voltage during brownouts to prevent the batteries from being required for anything less than a total power outage. The fact that the UPS is interacting with the electricity also allows it to use some of the power to continually charge the batteries and to keep them at capacity for when you need them.

Gas- or diesel-powered generators can create electricity, but it is extremely *dirty* (possessing a wide and erratic range of voltages). Your computer hardware can't handle directly connecting into a generator for any length of time. The quantity of surges, spikes, and brownouts coming from the generator takes a toll on the delicate circuitry of a computer server. If you must use a generator, filter the electricity through an online UPS.

Reinstalling Your Asterisk Source Code

The philosophical maxim of "If it isn't broken, don't fix it" is a great rule to work from with your Asterisk software. You shouldn't have to upgrade it or modify it during its lifetime.

You should reinstall your software only in the following circumstances:

- ✔ The new source code includes a feature that is beneficial to you.
- ✔ The new source code contains a fix to a bug that you have on your system.
- ✔ Your carrier uses the new source code that is causing conflict in your transmissions.

Making a backup copy of your original Asterisk code

Back up and remove all the files in the following directory prior to installing a new source code of Asterisk:

```
/usr/lib/asterisk/modules/
```

Each source code of Asterisk can have its own unique set of modules. Backing up the files and temporarily moving them to another directory keeps them safe while you are installing the new version of the source code. It doesn't matter where you save the files; just don't forget where you saved them.

All the Asterisk *modules* (shared objects) are compiled against the Linux kernel. You might not have to recompile Asterisk itself, even though it's a good idea, but we recommend recompiling the libpri and Zaptel drivers (kernel drivers) for the physical cards in the server.

You can copy the files for safekeeping with this command:

```
cp <source file(s)> <destination directory/filename>
```

The `cp` command tells Linux to copy something. The rest of the code identifies the source file or files and the destination directory that is receiving the copy.

This step establishes the copies as a fail-safe. If something goes wrong, bring the files back to their original location and all is forgiven. If your new Asterisk source code is compiled successfully, you don't need to worry about the copies you made. The `make install` command reestablishes the relationship between these files and the new version of Asterisk, so nothing else is required of you.

After backing up the applications and moving them to a safe place, back up your license directory, which can be found in the following directory:

```
/var/lib/asterisk/licenses/
```

The license directory holds all the G.729 VoIP codec licenses you have for your Asterisk server. Asterisk is an ongoing development in a constant state of flux, so adding new source code could potentially wipe out this file. Always back up this file and save it off-site with the rest of your important data.

Now that your applications and licenses are backed up and saved in a secure place on the server, take a quick inventory of your device drivers. If you have third-party drivers on your server with which Asterisk must interact, you may have to recompile them.

See Chapter 2 to find out how to recompile the Asterisk software.

Dealing with Linux ramifications

The relationship between Linux and Asterisk is important to understand because it also sets the stage for the relationship between Asterisk and its supporting drivers and software, such as libpri.

The *kernel* is the basic element of Linux. It is the DNA of the creature that weaves the fabric on which the rest of the software on the server is integrated. The other compiled software could be binary, modules, or compiled objects that all use a part of the Linux software to run. Without the support of the Linux environment, the other software can't function. As a result, whenever you change the Linux kernel, you must reestablish the interdependent relationship with the Linux operating system. Reintroducing the binary and modules links them to the new Linux kernel and allows them to function in the environment.

You can have two different versions of Linux running on the same server. It isn't common, but the spilt personality can effectively exist. Specific features may not be common to both versions of Linux, and each may be useful in a way that you want to keep it on the system. You can select the specific version of Linux you want to run by using the boot loader to differentiate each kernel of Linux.

A *boot loader* (sometimes called a *bootstrap loader*) is a small but important piece of software that bridges your hardware to your software and allows it to boot. The boot loader sometimes needs help and uses another boot loader, called a *second-stage boot loader,* to make the required connection to the operating system software. The second-stage boot loaders for Linux are called LILO or GRUB. Each loader has a menu file that contains specifics about the functioning operating system (OS).

Boot loaders exist to bridge the initial gap between your server and your operating system software. Hardware, such as servers or PCs, can only run software or execute commands that exist in their RAM or ROM. The challenge is that operating system software is commonly stored on the hard drive or a floppy disk. To access the operating system software, the hardware uses the boot loader to initiate the connection to the software, allowing the system to come to life and the hard drives to be accessed for the functionality provided by the software.

Providing General Server Maintenance

The level of software maintenance that your Asterisk server requires depends on its level of activity. Just as a car that is only driven 5,000 miles a year doesn't need an oil change ever other month, if you use an Asterisk server solely as a phone system, it doesn't require much housekeeping.

Asterisk isn't a messy program. It doesn't generate little piles of data debris in day-to-day operations. Depending on the amount of traffic pumping through the system, you may only need to clean the system by defragging it once a year. (*Defragging,* or *defragmenting,* the Asterisk's hard drives returns the files on the drives to contiguous clusters.)

 Start the defragmenting process at the end of the day to prevent the process from bumping into other active programs. When you return the next day, your server will seem like it spent a day at the spa.

Adding Organization to Your Dialplan

Dialplans are similar to local-area networks (LANs) because you should do some planning before you build them. That may be the ideal, but the reality is that dialplans generally grow organically based on the requirements of your application than methodically from a structured master plan. Taking the time for some general planning prevents massive reworking of your code later on.

Here's how to do it diagram your dialplan:

1. **Answer the question "What do I want the system to do?"**

   Are you building your Asterisk to provide service to calling-card customers, or is its primary function to receive and transmit calls for your office?

2. **Determine what it takes to process the calls and what supporting features are necessary.**

3. **Adjust the code to streamline it.**

   Your dialplan grows as you add more users, voice mailboxes, extensions, and devices, so you may find places where it's gotten out of hand.

4. **Replace any section of code you find yourself typing repeatedly with a macro.**

   The best way to ensure consistency is through the use of macros. Not only does the macro eliminate the possibility of typos, but it also gives you greater flexibility for changes. If you add, remove, or rename an extension, changing the one reference to it in a macro is easier than finding and changing each occurrence in the code.

Designing a Disaster Recovery Plan

The exact disaster recovery plan you build should be based on the level of availability required of your server. If you are using the Asterisk for your office, you won't be using the server after business hours. On the other hand, if your Asterisk server is being used as an international calling-card platform, you may have a constant flow of calls coming into your Asterisk at every hour of the day and night.

If your system must be up 24 hours a day, 7 days a week, you need the following as part of a disaster recovery plan:

- **Battery backups:** These range from standby UPSs to full disaster recovery systems with generators and online UPSs.

- **Surge protectors:** Your hardware is damaged when it receives a spike of electricity. Not only should you protect your UPSs with surge protectors or power conditioners, but if you are in an area prone to lightning, you should also consider adding protectors for your phone lines. Any cable over six feet in length attracts transient voltage in the atmosphere. Whether someone is arc welding in the adjacent office or you are in a geographic area known for electrical storms, the danger is the same.

- **Redundant Array of Independent Disks (RAID) hard drives:** A *RAID hard drive* configuration takes two hard drives of equal size and makeup and ensures that all information on one hard drive is also populated on the second hard drive. The philosophy is to create a fail-safe drive. If one drive fails, the second drive takes over without missing a beat. A multitude of configurations are available for your RAID hard drive.

   We recommend using a RAID configuration with mirroring (RAID 1). This provides you the redundancy to continue working, without losing data, in the event one hard drive crashes. You can install RAID 1 sets that consist of multiple mirrors, or RAID levels 0 and 1, but for a standard Asterisk configuration, these options are overkill.

Check out the following Web site for more information on RAID:

```
www.pcguide.com/ref/hdd/perf/raid/levels/index.htm
```

- ✔ **An active backup server that keeps a real-time copy of your Asterisk configuration files.** A real-time copy prevents you from losing any files, including the extensions.conf file.

- ✔ **Nightly tape backups of the hard drive that are stored off-site:** Many capable manufacturers of tape backups exist. Check out www.cdw.com and www.tigerdirect.com for information and pricing on the latest systems. Depending on your application and budget, you may opt for a Seagate, a Quantum, or maybe an ExaByte. Research the market and make your decision. If you are a carrier, you may have to use more than one tape to store all your data. Fortunately, most tape backups allow your data to span more than one tape. After the tape backup is complete, take the tapes home or put them in a safe place. Keep and rotate through three sets of tapes: One set stays at your safe location, a second set is used for your current backup, and the third set is a spare.

 If you have sufficient funds to spend on disaster recovery and your application is valuable enough, consider replicating all your active production software and keeping it off-site.

Securing the System

Security is an issue for any network; the fact that you're running an Asterisk server doesn't make it more or less prone to attack. Checking the log files is the simplest way to see whether your server has been compromised.

Use the vi or view command in Linux to inspect your log files. You don't have a special program or filter that only displays errors or aberrant programs active on your server, so you'll have to do some detective work. The log files are typically stored in this directory:

```
/var/log/<programname>/logfilename
```

Execute the following command to view the current file for the Linux kernel:

```
vi/var/log/kernel/current
```

The `view` command is in the same position in the syntax of the code. Begin the line of code with the command, insert one space, and list the filename you wish to view.

If you don't recognize some of the programs running in the log files, investigate them. This is the easiest way to scan your server for aberrant activity. This test also reveals that your Asterisk software is not alone. Any program you have may help someone compromise your system. Ensure that all software running on your server has the latest security patches.

Ip-tables is advanced firewall/filter/routing software that can protect your server. The software is a bit of a challenge to put together. You can filter incoming packet data traffic by port, packet inspection, and IP address, for starters. It also handles the processing of outgoing packets from internal servers. It requires some advanced study before configuration and installation. You can find more detailed information on ip-tables at the following Web site:

```
www.cae.wisc.edu/site/public/?title=liniptables
```

Part IV
The Part of Tens

"I tried calling in sick yesterday, but I was put on hold and disconnected."

In this part . . .

We cover the ten things you should never do — the cardinal sins of Asterisk — in Chapter 16. Read this chapter carefully and pay attention.

The middle chapter (Chapter 17) in this part is much more Buddhist-friendly. We provide a list of ten things we encourage you to do.

Finally, we remove any remaining anxiety you have ever had about Asterisk by giving you ten places to go for help (Chapter 18). Enjoy these words of wit and wisdom; they are the best we have.

Chapter 16

Ten Things You Should Never Do with Asterisk

Never is a very absolute term. It doesn't leave much room for negotiation. Our world isn't so black and white as to make anything a *never*, but it is easier to call a chapter "Ten Things You Should Never Do" than "Ten Things That Aren't Preferable to Do, so Take It under Advisement and Try to Avoid Them."

That being said, we offer you Asterisk things to avoid. Engaging in these activities can result in legal action or programming frustration.

Killing Your Carrier with Calls

Every carrier loves to have a lot of traffic. The more calls, the more minutes, the larger your phone bill, the more money the carrier makes. But you can have too much of a good thing. Setting up a cluster of six Asterisk servers to dial ten calls a second each to a small town in Iowa does bad things to the telecom world. Here are two reasons why you shouldn't do this:

- ✔ Slamming your carrier's network with 60 calls per second (hey, that's 3,600 calls per minute) can preoccupy it so much that it may lose the ability to provide service to other customers.

- ✔ You can't continually hit your carrier with 60 calls per second because eventually you fill up your outbound capacity. Then, as the calls are answered and connected, conversations take place, and the call is disconnected, the dialing process begins again.

Some people obsess about the calls per second possible on outbound calling. It all seems like a bizarre desire for the fastest 0-to-60 time. The reality of it is that eventually the 10 to 30 seconds of time where the call is ringing and the duration of time you are speaking on the phone negate a real need for extremes of call processing in excess of one call per second per T-1 digital circuit.

A *T-1* line is the American standard for digital dedicated service and is sometimes also called a *DS-1* line. Each T-1 consists of 1.544 Mbps of bandwidth that can be multiplexed into 24 individual channels, each called a *DS-0* and each being capable of handling one phone conversation. All dedicated circuits are established on the basis of T-1s. For example, a DS-3 is a circuit that contains 28 T-1s. The terms *DS-0, DS-1,* and *DS-3* are used to identify the individual call-level circuit (DS-0), the first-level aggregated digital circuit (DS-1), and 28 of those DS-1s (DS-3). The only difference is that in other countries, you get more DS-0s in your DS-1s.

Carriers have the following volume constraints:

- ✔ **Incoming facilities:** The non-VoIP world has large telephony switches with cards in them, similar in function to the Zaptel cards in your Asterisk server. Depending on the type of switch your carrier is using, the card may handle 28 to 128 or more T-1s of bandwidth. The cards are housed in racks in the switch, and this is where it can get dicey. For each rack in your carrier's switch, one tone generator exists. The *tone generator* is a device that provides dial tone to every DS-0 housed on every T-1 on the rack. If every DS-0 attempts to process a call at the same second, the tone generator becomes overwhelmed, and some of the attempted calls don't receive dial tone. These calls without dial tone fail, and then people get upset.

   If you are causing disruption to your carrier's network at this level, the carrier may ask you to scale back your dialing, or it may just restrict you to a certain quantity of calls per second and not even tell you.

✔ **Egress facilities:** Even if your carrier receives your calls, they still have to get through the network and be deposited to the *inbound tandem,* essentially a one-way road into the local carrier that provides the dial tone for the number you dialed. The problem is that long-distance carriers are business-minded entities. They don't like to have spare capacity lying around. Every T-1 of connectivity to every local carrier to which they terminate calls costs them money every month. If the long-distance carrier has a consistent volume of three or four T-1s of bandwidth to a certain location every month, the long-distance carrier may allocate one or two more T-1s of bandwidth for an unexpected spike in traffic. If you are dialing to a small end office in Red Oak, Iowa, and trying to ring all 6,000 people there within 5 minutes, you will clog all traffic going into that town for as long as you hammer the switch. So don't do that.

In this instance, the local carrier identifies the long-distance carrier as jamming its switch. The long-distance carrier identifies you as the source, and then your service is restricted to a specific volume of calls per second or per minute — or your carrier just blocks your service.

Manipulating the Origination Phone Number

Don't use your Asterisk to manipulate origination caller ID phone numbers. The cost for placing a call to one state from another (interstate call) is almost always cheaper than the rate charged for calling within the same state (intrastate call).

The rate charged on calls is based on comparing the origination phone number to the number dialed. The local carrier completing the call identifies your call as intrastate and bills the long-distance carrier at the intrastate rate. The long-distance carrier, in turn, charges you the intrastate rate.

The way to a cheaper long-distance bill seems clear: Pass an out-of-state origination phone number for every call, and you can reduce your cost considerably. Intrastate rates can reach 15 to 17 cents per minute, while interstate rates might be as low as 3 cents per minute. It is a large incentive, but don't do it.

Manipulating the origination phone number with the intent of getting the lower interstate rate is fraud. The Federal Communications Commission (FCC) can initiate legal action against you, and you may pay fines or even serve jail time.

Manipulating the Telemarketing Caller ID

As legally dangerous as it is to manipulate the caller ID on outbound calls, it's also potentially expensive to send the correct caller ID. This is a larger concern for companies using Asterisk to send calls for telemarketing campaigns, but it is something to keep in mind.

The FCC has caller ID requirements for all telemarketers that can expose them to intrastate rates in states that they aren't calling from. The rules stipulate the following on caller ID presented by telemarketers:

✔ The caller ID must be a legitimate phone number.

✔ The number provided must be able to receive calls from people who wish to remove their name from the telemarketer's calling list.

Many telemarketing companies don't market for themselves, and simply run calling campaigns for large retail or sales companies. Each of their clients may require that the caller ID presented is a local phone number to the area they are marketing. This is frequently used because people receiving local calls are more likely to answer the phone than people receiving anonymous calls from distant companies. Presenting a callback number for the local retail store also allows the telemarketing companies to avoid irate calls from people who don't want to be called.

The problem is that carriers frequently rate your call as either inter- or intra-state based on the origination caller ID presented and the phone number dialed. It doesn't matter if the call actually originates in Texas; if your caller identifies a South Carolina phone number as your caller ID, and the number you're dialing is also in South Carolina, you're charged the more expensive South Carolina intra-state rate. Depending on the long-distance carrier you're using, and their transmission and signaling options, you may have a way to avoid the higher rates. If they don't have a work-around, be sure to factor in the higher rates.

Forgetting to Build Routes

Don't forget to build new phone numbers and Direct Inward Dial (DID) lines into your system as soon as you know the number. If you are migrating phone numbers with the Local Number Portability (LNP) process from one local phone carrier to another, build the routes for the number as soon as you start the LNP process.

Many companies prefer to wait out the LNP process and only install the numbers on the morning of the date they are to be released to the new carrier. We are all for conserving energy. Heck, why build routes for numbers that aren't even active and might not be for weeks, or possibly months? The answer comes in the form of a 2:00 a.m. wake-up call, when one of the numbers ports early or moves so fast through the process that by the time you get an update on the number, it has moved. If you don't have the numbers programmed in the Asterisk, the calls being sent to you fail and the number is dead.

Getting an Edge on Radio Call-Ins

Programming the phone number of your favorite radio show may be helpful, but having Asterisk send out one call a second to the show to tie up the lines is unfair. You do increase your chances of being "caller 12 to win the tickets," but until everyone has Asterisk to dial out multiple calls per second, it isn't a level playing ground.

If you must do something like this — we're sure you have a good reason — the simplest way is to build a loop into your dialplan like this:

```
exten => 2565551212,1,Dial(Zap/g0/${EXTEN}|20)
exten => 2565551212,2,Goto(1)
```

The two lines of code take your call to 256-555-1212, dial it out on a Zaptel channel, and wait for 20 seconds (if the call doesn't receive an immediate busy or a connection). After 20 seconds, the dialplan cycles to priority 2, where the call is sent back to priority 1 and is attempted again. The cycle continues on this dialplan until you hang up.

This scenario simply initiates and reinitiates an outbound call to a specified phone number.

If you want to create multiple calls to the same phone number, things get more complicated. Start your inquiry with the Asterisk *spool directory* that's used to queue and process calls. Placing a program file into the spool directory that defines the parameters of the call allows Asterisk to dial the phone number as quickly, as often, and as many times as you need. After the person answers your call, you can program Asterisk to send the call to your extension, your voice mail, a queue, or wherever you deem appropriate.

Exposing Yourself to Fraud

Tightening your dialplan is one of the best ways to prevent your system from being used to make fraudulent calls. The most typical, and most costly, fraud involves a Call-Sell operation. *Call-Sell operations* are generally set up to use your phone system to sell people an unlimited call to anywhere in the world for a set fee, maybe $20. Businesses with large dedicated circuits can easily suffer $50,000 or more over the course of a weekend in fraudulent charges.

The most common way that the Call-Sell operations make the calls is by accessing your phone system with a toll-free number and then hacking around to determine what is necessary to find an outside dial tone. When the hacker believes he has found a dial tone, he makes a test call to an international country and then starts selling your phone service.

Fortunately for you, Asterisk is an intelligent system. You should prevent an anonymous caller from accessing outbound dial tone by restricting that privilege to only select calls from specific phone lines. The caller ID received on an incoming call can be built to check against a database of individuals allowed to make outbound calls. If a caller ID isn't on the list, that caller doesn't get an outside line.

The following code shows how you can qualify a call in your dialplan. Only calls with the correct origination phone number are authorized to make an outbound call and are sent to the initial extension of the [outbound-available] context. All other inbound calls are immediately sent to the [outbound-unavailable] context.

```
[default]

exten => 2565551212/2567142943,1,Goto(outbound-
          available,s,1)
exten => 2565551212,1,Goto(outbound-unavailable,s,1)

[outbound-available]

exten => s,1,Background(enter-the-number)
exten => s,2,WaitExten
exten => _NXXNXXXXXX,1,Goto(outbound,${EXTEN},1)

[outbound-unavailable]

exten => s,1,Playback(calling-unavailable)
exten => s,2,Hangup
```

Building a Dialplan of Frustration

As powerful as Asterisk is, it is powerless against the unseasoned programmer. Every priority of every dialplan must be well thought out. A minor programming mistake in one place can bring down the entire system and stop all communications through the server.

One of the most dangerous pieces of code in the Asterisk lexicon is this:

```
_X.
```

We know, it doesn't look like much. Heck, it's only three keystrokes. How bad could that be? Well, really bad. Those three little keystrokes represent a code for pattern matching that applies to every phone number in the world. Placing that code within your dialplan for incoming calls captures every call and routes it as you have specified. You can easily end up sending every inbound call to your Asterisk to oblivion or to an automatic hangup. This isn't nearly as painful as dropping it into your [outbound] context. Depending on your call volume and type, a few hours may elapse before you realize you have a problem. Then you may need a few more hours before you find the small glitch in the code.

Working without a Safety Net

Brownouts, blackouts, and power surges are no joke. Any of these problems could spell disaster to your Linux server running Asterisk. If your business is in an area know for electrical storms, like the lightning alley on the border of Florida and Georgia, ignoring the reality of electrical surges could prove fatal to your hardware. Any cable longer than six feet attracts the transient voltage released during a lighting strike. If your copper digital T-1 line runs from one end of your building to another, it might as well be a lightning rod.

Purchase inline surge protectors for your analog and digital circuits, as well as building a line of defense with Uninterruptible Power Supplies (UPS). It's always better to lose an inexpensive UPS that doesn't contain 100 hours or more of code you've input for dialplans than the Linux server sitting behind it.

Nothing can save you from a direct lightning strike, but we've only heard of that happening once to a long-distance carrier's switch in Palmdale, California. One day it was a lively switch, processing millions of calls, and the next day it was a smoldering, melted, contorted piece of silicon and steel modern art.

Disregarding the Power of Macros

Macros are a gift to programmers, so use them. The likelihood that a repetitive section of code is mistyped increases every time you type it. Don't forget to use macros whenever you can.

Macros are similar to functions or methods in traditional programming. Imagine that you have 100 extensions on your server to be programmed. Each extension consists of code that performs the following actions:

1. **Dial the extension.**

2. **If answered, hang up.**

3. **If unanswered, identify the reason and route accordingly:**

    • **Busy:** Voice mail

    • **No answer:** Voice mail

    • **Congestion:** Play unavailable message

The simple section of the dialplan for each extension may take ten lines of code. Without a macro, you type 1,000 lines of programming. With a macro, you write the program once and reference it on each extension with a single line of code.

Even after you write the macro for each extension, the benefits of macros continue. Modifying all the extensions to integrate queuing or redirecting calls for voice mail is efficiently done in the macro. You make the fix once and allow it to affect all 100 extensions, instead of having to type each one individually.

Ignoring the Need to Learn More

If you have been in telecom for years handling voice issues, the programming and data aspects of Asterisk may be foreign to you. Don't read this book and think you know everything necessary to administer your Linux server, maintain your Asterisk server, and troubleshoot VoIP. Make it your mission to find out more about Linux, Asterisk, VoIP, and general programming if you are going to be responsible for keeping the Asterisk server working.

If you aren't going to take on the responsibility of maintaining the Asterisk system yourself, hire someone who is competent to do the job right. Ensure that he or she has sufficient programming and Linux knowledge to properly handle the tasks at hand, and then push the person to continue to gain knowledge. As knowledge of these systems grows, so does the flexibility available to you through Asterisk.

We recommend any of these *For Dummies* books if you need to brush up on your programming:

- *UNIX For Dummies* by John Levine and Margaret Levine Young
- *Python For Dummies* by Stef Maruch and Aahz Maruch
- *PHP For Dummies* by Janet Valade
- *PHP & MySQL For Dummies* by Janet Valade
- *Linux For Dummies,* 5th Edition by Dee-Ann LeBlanc
- *Red Hat Enterprise Linux 4 For Dummies* by Terry Collings
- *C For Dummies* by Dan Gookin
- *C# 2005 For Dummies* by Stephen Randy Davis and Chuck Sphar

Chapter 17

Ten Fun Things to Do with Your Asterisk

Asterisk is a workhorse with so much to offer businesses that you may forget to take the time to do some fun things. The majority of this book tells you how to make things work, troubleshoot, install, and build functionality. That is helpful, but all work and no play can make you a dull person.

This chapter is about using Asterisk for fun and quirky things — simple things like routing telemarketers down a dialplan gofer hole. Granted, it isn't the most efficient use of programming time, but it sure is fun. Read through this chapter, relax, and soak it all up.

Fending Off Telemarketers

You can fend off telemarketers with Asterisk. Federal Communications Commission (FCC) regulations state that every telemarketer must provide a valid caller ID number on every call. A caller ID number allows you to receive the call and route it based on the number. If a certain telemarketer has called you consistently and won't take you off its list, you can identify its caller ID number and send its calls to an endless loop. Any recording in a GSM or WAV format does the trick. Here are some we recommend:

- ✔ The Hamster dance
- ✔ A book on tape converted to a WAV file
- ✔ Barney, the big purple dinosaur

Save the file in the /var/lib/asterisk/sounds directory. Name it something simple and easy to remember because you need to input it into your dialplan.

You don't have to place the file in the sounds directory or a subdirectory of it. We simply suggest you keep it there to make it easier to find later when you want to replace it with something newer or more monotonous.

The line of code you add to your dialplan looks like this:

```
exten => 2565551212/2567142943,1,Playback(custom/telemarketer)
```

The Playback() application is your main ally in this line of code. It instructs Asterisk to play the telemarketer file found in the /var/lib/asterisk/sounds/custom/ directory. You don't have to list the extension of the file in the code. Asterisk takes all sound files named telemarketer, regardless whether they are WAV, GSM, or something else. Asterisk plays the file that's easiest to convert based on the current channel format.

Giving Your Friends Special Options

You can also use your power to build a queue into your dialplan for all your personal friends and relatives based on their phone numbers. Give your friends the options to speak to you directly or to access any of your WAV files. Setting up an elaborate series of prompts is a bit complex but can be worth it for you. You need to record each greeting and save it in a WAV or GSM format.

The dialplan code for this little farce isn't that complex. Start off with the s extension as you would for any queue, and begin by playing the message of your choice. The code looks like the following:

```
[context]
exten => s,1,Background(testmessage)
exten => s,1,WaitExten
exten => 10,1,Dial(SIP/${EXTEN})
exten => 12,1,Dial(SIP/${EXTEN})
exten => 17,1,Dial(SIP/${EXTEN})
```

In the case of our dialplan, we called the voice file testmessage. Replace _{EXTEN}_ with the specific SIP extension or other type of extension to which the call should route. To add additional layers to the dialplan, take the extension to which the call was sent when the caller presses 10, 12, or 17 and create another Background() message with another set of options. Layer them on until the game is taken to whatever illogical conclusion you can devise.

Setting an Extension to Call for Local Weather Reports

Asterisk is informative as well as helpful. The weather is always a topic of conversation, so why not integrate it into your Asterisk database? Wouldn't it be nice to dial up your local forecast just as easily as calling your receptionist? Go ahead and do it.

The following code establishes an extension to your [internal] context (in our case, extension 65) that plays the weather report:

```
#!/bin/bash /usr/bin/text2wave /tmp/weather.txt -F 8000 -o
            /tmp/weather.wav
chmod 755 /usr/bin/convert
```

The first line of the code tells Asterisk that the /bin/bash program is going to execute the script file. The second line of code, starting with chmod, creates a file called /usr/bin/convert that's accessed to run the program. Place this code in the /usr/bin/convert file.

After you establish an internal extension, you have to add the extension *65 to your [internal] context in your extensions.conf file. The code to make that happen is as follows:

```
exten => *65,1,Answer
exten => *65,2,Playback(national-weather-service)
exten => *65,3,System(/usr/bin/curl -s
        ftp://weather.noaa.gov/data/forecasts/city/ne/o
        maha.txt > /tmp/weather.txt)
exten => *65,4,Wait(1)
exten => *65,5,System(/usr/bin/convert)
exten => *65,6,Playback(/tmp/weather)
exten => *65,7,System(rm /tmp/weather.* -f)
exten => *65,8,Hangup
```

If you aren't in Omaha, NE, you need to change the city and two-letter state abbreviation in priority 3 to those of your hometown.

This fun little section of code retrieves the local forecast from the National Weather Service in text format. Asterisk converts the text file to an audio file and plays it to you when accessing extension *65 within your Asterisk. Now that's cool!

Making this happen takes some manipulation of files within Asterisk. If you don't know how to move around your Linux server and modify files, check out Appendix C for help.

Creating a Voice Mailbox That E-Mails Everyone

If you are expecting a baby, about to get engaged, or having a wonderful life-altering experience, tell people about it. Set up a voice mailbox with all their e-mail addresses so that you simply have to call one number and leave a message, and your Asterisk sends the message to everyone. You have to work a little Asterisk magic behind the scenes, but it is worth it.

You could even use this feature to send blast messages for business purposes. When you release a new product or sales promotion, leave a voice mail on the AllSales voice mailbox and have it sent to all your sales staff.

You can send voice mails to a group of e-mail addresses by creating an alias e-mail account under your normal sendmail setup located in the /etc/mail/ aliases directory. To set up the alias group, simply separate the alias (in this case Family) from the individuals with a colon and separate the individuals from each other with a comma. For a group of relatives, it may appear like this:

```
Family: Momma@aol.com, Brother@yahoo.com, Uncle@cox.com
```

 If you don't know how to edit files, check out Appendix C. Knowing how to edit files is a useful skill when working with Asterisk.

Using the Power of VoIP from Your Cell Phone

If you find yourself on the road and don't have your cellphone or calling card, you can set up a dialplan that allows you to call into your VoIP switch, authenticate yourself, and then call anywhere else.

You can configure this functionality for whatever your application demands. You can set up a toll-free number to ring into your Asterisk switch, and after authentication, allow you to dial anywhere in the world. You can also configure your Asterisk to receive a local call and give the call outbound dialing access.

```
[inbound] ;[default]
exten => 2565551212,1,Answer
exten => 2565551212,2,Wait(2)
exten => 2565551212,3,Authenticate(52467)
exten => 2565551212,4,Goto(ontheroad,s,1)

[ontheroad]
exten => s,1,Playback(enter-phone-number10)
exten => s,2,WaitExten
exten => _NXXNXXXXXX,1,Goto(outbound,${EXTEN},1)

[outbound]
exten => _NXXNXXXXXX,1,Dial(${TRUNK}${EXTEN})
exten => _NXXNXXXXXX,2,Busy
exten => _NXXNXXXXXX,102,Congestion
```

You start with the incoming call hitting your default [inbound] context where you enter your authentication code. The correct authentication code sends you to the first priority of the s extension, which gives you a dial tone to input the phone number you're dialing. Then after you input the phone number, you go to the [outbound] context to have your call sent over a carrier to its end destination, anywhere in the United States.

This system is also the basis of how calling cards work, but their dialplans are more complex. Not only must they receive calls and allow you to dial out somewhere, but they must also track your usage to manage the accounting of your per-minute charges and your available balance on the card.

If you allow the functionality whereby inbound callers can reoriginate dial tone and make outbound calls, build in a mechanism to monitor these calls. This is the primary means by which thieves hack your phone system and rack up tens of thousands of dollars in long-distance charges. If you are going to open this application, check the Call Detail Record (CDR) for the calls made on it every few days. It is better to be vigilant than to be surprised by a $68,000 phone bill at the end of the month.

Turning On Your Lights with Asterisk

Pushing the limits of Asterisk requires a bit more research. One of the most cutting-edge software applications available at this time is a programming language called X10. It works well in a Linux environment and allows you to manipulate electric devices in your home.

The setup is pretty straightforward. Replace some of your electrical outlets with either X10 receiving electrical outlets or external receivers. Then plug in lamps, appliances, and anything else you want to remotely control. A single device (a *transceiver*) that's plugged into an electrical outlet activates the receivers.

The transceiver is programmed to allow remote access to the basic functionality of the electric devices connected to the receivers. For example, you can remotely turn on the lights in your house or turn off an electric stove from your phone.

This functionality is accessible by integration with Asterisk in the dialplan. First you build authentication in the dialplan, and then you provide options. Turning on the light in the driveway may be press 5. Turning off the driveway light may be press 6. Turning off the stove off may be press 8, and turning on the coffee maker may be press 22.

This may require some additional homework, but the Asterisk software has a system application that allows it to pass command-line data to be executed. You may have to create individual contexts in the dialplan for the X10 connection, but if you have the time, it is a cool hobby.

Most modern homes are built with two 110-volt circuits in the breaker box. These circuits break down the 220 volts that are supplied from your power company to your home. Some of your electrical outlets are on one circuit, and some are on the other. When using X10 technology in your home, you usually need a *phase coupler* to connect the two 110-volt circuits. The phase coupler facilitates X10 communications throughout your home.

Remote Listening

New parents generally have the jitters when leaving their baby for the first time. Is the babysitter good? Is she going to throw a party while you are out? Is your baby going to cry all night, and is the sitter going to ignore him? Well, you can set up your Asterisk to tap into the baby monitor at your house.

You need a few things to make this work, but it's pretty straightforward. You must have these essentials:

- ✓ **An Asterisk server at your home:** Maybe it's the lab machine you toy with on the weekends, or you may be using it for fun as your home phone system. The bottom line is that your Asterisk must be physically located close to the baby monitor.

- ✓ **A baby monitor with an output jack:** Many baby monitors come equipped with small bantam jacks used to connect headphones or speakers. This is how the transmission is sent from the monitor to Asterisk.

- ✓ **A sound card in the Linux server on which you're running your Asterisk:** This is how you receive the transmission from the baby monitor.

- ✓ **A cable to connect the baby monitor to the server:** You need to connect the jack on the baby monitor to the interface on the sound card in your Linux server.

Wiring the baby monitor and the Asterisk box together completes the physical connection. Build the port on the sound card into your dialplan by programming it to dial your cellphone. After the call is initiated and connected, you can listen to what is going on from anywhere in the world (assuming that you have global coverage on your cellphone).

Transmitting Your Voice through Your Stereo

This little trick works only if you have an Asterisk server installed at your house. The main requirement of converting your stereo system into a broadcasting network for your voice is wiring your stereo system into the sound card in your Asterisk server.

The premise behind this fun experiment is to wake your kids promptly at 8:00 a.m. when you're at work. Install a new audio card in the Asterisk box running at your home, and then run a cable from the audio card to your stereo. You may need to get a converter to allow the audio card to connect to the AUX1 or AUX 2 port on your stereo, but your local Wal-Mart or Radio Shack has everything you need to convert RCA jacks to SPDIF jacks — or whatever you need.

After the wiring is complete, configure the device as an extension to receive incoming calls. The only difference between this and a normal extension is that the extension is designed to use the dial() application to reach the sound card. For example:

```
dial(OSS/dsp)
dial(ALSA/default)
```

These sections of code represent a call being sent to either an OSS card with a dsp driver or an ALSA sound card with a default driver. Research the card to ensure that you have the correct driver listed and installed.

If you are really tricky, you can use your main office phone system to call the Asterisk server at your house that is connected to the stereo and blast your voice through your house remotely. After it is set up, you can configure any number of ways to access the system and broadcast your voice from your stereo. Have fun; get creative!

Maximizing Your Savings

Asterisk allows you to save money by using VoIP and providing access to multiple carriers for every call. If you're using standard telephony DS-1s right now, you're restricted to transmitting 24 phone calls at a time. Using the VoIP capabilities of Asterisk and a codec with compression allows you to double the amount of calls possible over the same bandwidth.

The multiple interfaces of digital, analog, and VoIP, each with their potential to connect to a different carrier, allows your second wave of cost savings. Every carrier connected to your Asterisk establishes a unique pricing structure for your company. Compare the pricing plans to find the lowest rates and integrate them into your dialplan. Pattern matching (covered in Chapter 5) allows you to identify calls and route them to a specific carrier based on the area code dialed, or even more specific. The more money you spend on long-distance, the more you can save with this type of Least Cost Routing (LCR) setup.

Taking Charge of Your Phone System

Asterisk is built by your hands and for your specific application. Don't be the victim of a proprietary phone system ever again. Don't suffer with features you don't need, or features you want but can't have. Take responsibility for all of it and build it just like you need it.

The open-source nature of Asterisk empowers you to create it just as you need it, and customize it to reflect who you are. You can't build a blacklist of unwanted calls to send to a 20-minute recording of the Hamster Dance on any other system, so customize it and make it yours.

Chapter 18

Ten Places to Go for Help

*T*his book provides a great introduction to Asterisk, Linux, and some other helpful software, but you may need something more. Over time, your need for troubleshooting help diminishes and your need for programming advice grows. This expected learning curve may leave you frustrated at times, but help is available.

Asterisk has a supportive community of developers and programmers. They've built many Web sites to assist you with advanced Asterisk questions and to work you through some of the basics. We're covering our favorite places to go for help in this chapter. We're sure that they will be as helpful for you as they were for us.

Working the Asterisk Wiki

A *wiki* is the technological equivalent of a co-op. It is the site where people come together to help each other and work on projects. It isn't some over-sized barn with a corrugated metal roof like the old food co-ops of the 1970s; it's actually a Web site. This is our favorite wiki for Asterisk issues:

```
www.voip-info.org/wiki
```

This wiki not only posts job offerings for Asterisk technicians, but it's also invariably the first place on the Web where you'll search for help with Asterisk.

Going to the Source

If you have a huge application issue you need to sort out, you can hire one of the authors to help you. Brady Kirby is the president of Atlas VoIP, and he can be found on the Web at www.atlasvoip.com. His company has been working with Asterisk for years and is working on some amazing software to help realize the true potential of our favorite open-source system.

Atlas VoIP focuses its development efforts on the following two most important Asterisk concerns:

- **A Web-based portal for Asterisk customers to use, allowing them real-time access to add and change aspects of their customer account:** You can make changes to active IP ports and their configurations, and you can add, change, or delete individual or groups of phone numbers.

- **Integrating multiple servers to process calls:** The scalability of Asterisk is greatly hindered because each server functions independently. You can't load Asterisk on multiple Linux servers and have them all share dialplan information without building your own software. With that in mind, Atlas VoIP is building the software for you. Atlas VoIP is constantly polishing it for better performance, but the end result allows you to harness the power of multiple servers.

Developing Your Database Skills

Asterisk has great software for what it's required to do. Some limitations require database skills. The ability to parse your Call Detail Records or share information between Asterisk servers is integral to using the software to its fullest. Third-party software may be of some use, but it isn't going to be designed around your business application.

Fine-tuning the reports available and knowing how your Asterisk servers interact fall to a person with the skills to bring it all together in a database. The best place to start is with MySQL.

The MySQL Web site at www.mysql.com is loaded with everything you need to get comfortable with the software. MySQL offers a training program to bring you up to speed. To talk to someone directly, you can call MySQL from the United States toll-free at 1-866-697-7522; from international locations, call 1-208-514-4780.

Strengthening Programming Skills

The basic environment of Asterisk dialplans is more closely related to computer programming than anything in telephony. The software takes basic telephony and packages it using programming. The toggling of 1s and 0s in traditional telephony is built into the soul of Asterisk. The basic features of multiplexers and phone systems are now applications available within the programming of a dialplan.

The challenge is that people who have worked on voice circuits and systems aren't necessarily programmers. You find two types of technicians in telecom: the data folks and the voice folks. Data technicians, with experience building Frame Relay, Asynchronous Transfer Mode (ATM), and Multiprotocol Label Switching (MPLS) networks, may be more familiar with the basic tenets of programming, but to the voice folks, this can be a whole new world.

If you haven't worked as a programmer and you aren't going to outsource the development and maintenance of your Asterisk, you need to understand basic programming. Most city colleges offer basic programming courses. If you don't have a consistent schedule and can't make it to a city college, take an online course. We recommend starting with C/C++ because it's a good general programming language. Then you can branch out to Perl and Python.

C/C++, Perl, and Python may not have a direct benefit to your Asterisk dialplans. But if you understand how the software is constructed and how it executes tasks and transfers information, you can develop an ability to visualize how the software works. Viewing the tasks in your mind's eye that must occur in a given sequence within a program can give you perspective. This evolved perspective enables you to understand all the required elements necessary to bring an application to life.

In this respect, understanding programming is similar to learning a spoken language. You first have to learn the individual words and the rules of grammar. Only after you understand these elements can you put the words together in a logical manner to convey information in sentences. This ability to bring the required information together allows you to create dialplans in Asterisk.

Tapping All Known Resources

If an issue is over your head, don't be afraid to ask for help. Call or e-mail the company you purchased your hardware from, someone you spoke to at a trade show, or as a last resort, your carriers. These resources might know how to program, configure, and design your Asterisk network, but every challenge you face won't be at that scale.

Asterisk is popular software, and everyone in the industry seems to know an Asterisk guru. Your friends and coworkers in the industry are a great resource for Asterisk programmer leads. These individuals generally live close to you and won't charge you for a 30-minute chat on how to streamline your dialplan. Even if they can't give you the specifics about integrating a six-server cluster, they can provide another link in the chain of your Asterisk education.

Loading Up on Linux

Asterisk lives within Linux; it's the very water in which Asterisk swims. Linux colors the way that Asterisk reacts and processes data and provides the stability to make it all work. Because so much of the administration of Asterisk depends on Linux, you need to become proficient in Linux.

The Linux Web site, www.linux.org, is a great place to gather general Linux information. The site offers information on the many distributions of Linux and provides documentation, applications, and the latest Linux news.

The Linux user forum allows users to exchange information and ideas. Each distribution of Linux, such as Red Hat or Gentoo, also has its own Web site and forums. They are helpful for specific issues particular to that distribution of Linux, but the forum at www.linux.org is the most active and widely used.

Entering the Digium Forum

Asterisk forums allow you to ask questions and get answers about your latest Asterisk challenge. The most helpful site for Asterisk users of all skill levels is http://forums.digium.com. The forums are grouped by category, and the information they contain runs the gamut from interesting trivia to basic information to solutions for large projects. The groupings are as follows:

- ✔ Users

- ✔ Developers

- ✔ Asterisk on BSD (UNIX)

- ✔ Asterisk on Sun — the Sun servers, not the celestial body

- ✔ Documentation

- ✔ DUNDi

- ✔ Jobs

You can also reach the forums through www.asterisk.org. Click the Support link, and then click the Forums link.

Sign up for the forums that interest you, and every day, a new e-mail arrives about the topic of your choice. The e-mails may be questions from people asking for help or answers to a question you've posted. If you are having a hard time deciding which forum to join, just sign up for the Users forum, which covers most of the basics.

The forums are open platforms for discussion, so response times may vary. A simple question worded diplomatically can receive a response in minutes, and more complex inquires may take days before anyone chimes in with suggestions. Our recipe for success is to always be gracious and ensure that you're posting your question on the right forum. Get a feeling for the forums before diving in with questions that may not be applicable to the group.

Part V
Appendixes

The 5th Wave   By Rich Tennant

" He seemed nice, but I could never connect
with someone who had a ring tone like his."

In this part . . .

Part II provides details on how to build individual sections of the dialplan, but it doesn't provide an overall view of the dialplan as a whole and how each section of it relates to the other elements. To prevent you from not being able to see the forest for the trees, we put the trees in Part II and the forest in Appendix A. Appendix A details the setup of an overall Asterisk dialplan.

Appendix B is a VoIP primer. It provides general information on the structure of a VoIP transmission. We've loaded this appendix with diagrams that allow you to see the constituent parts of a VoIP transmission. This allows you to easily understand how it works and exploit VoIP to the fullest.

Appendix C is a basic rundown of Linux. We know that many people with telecom experience in voice probably have no experience with Linux. So we've taken all the Linux bits and pieces you need to know to work with Asterisk and dropped them in Appendix C. So read and enjoy.

Appendix A

Visualizing the Dialplan

. . . .

In This Appendix

▶ Viewing the dialplan from 30,000 feet

▶ Reviewing contexts

▶ Understanding the purpose of each context

▶ Recording custom menus

. . . .

*I*n Part II, we show you how to build a basic dialplan with contexts, extensions, applications, and all the tools necessary to perform specific tasks. This appendix helps you flesh out your dialplan so that it functions exactly as you want it to.

Taking In the 30,000-Foot View

Before working on your dialplan, you need to remember a couple things. First, you must create a directory in Linux for your Asterisk sound files called custom. Locate the directory in the following folder:

```
/var/lib/asterisk/sounds/
```

Then you need to establish the structure that allows the background applications in the sample dialplan to function by executing the following two commands:

```
touch /var/lib/asterisk/sounds/custom/recording-menu.gsm
touch /var/lib/asterisk/sounds/custom/intro.gsm
```

The .gsm audio file can be empty, but you must reference it as a background application in the dialplan.

Listing A-1 shows a dialplan.

Listing A-1: A Sample Dialplan

```
[globals]                                                        →1
HOMEAC=256
TRUNK=SIP/111.222.333.444/                                       →3
[macro-stddid]                                                   →4
exten => s,1,Dial(${TRUNK}${ARG1})
exten => s,2,Goto(s-${DIALSTATUS},1)
exten => s-NOANSWER,1,Congestion
exten => s-NOANSWER,2,Hangup
exten => s-BUSY,1,Busy
exten => s-BUSY,2,Hangup
exten => s-CHANUNAVAIL,1,Congestion
exten => s-CHANUNAVAIL,2,Hangup
exten => s-CONGESTION,1,Congestion
exten => s-CONGESTION,2,Hangup
exten => s-ANSWER,1,Hangup
exten => _s-.,1,Congestion
exten => _s-.,2,Hangup                                           →17
[macro-stdexten]                                                 →18
exten => s,1(chanavail),ChanIsAvail(${ARG2})
exten => s,n,Goto(havechannel)
exten => s,chanavail+101,Goto(voicemail)
exten => s,n(havechannel),Dial(${ARG2},20)
exten => s,n,NoOp(${DIALSTATUS})
exten => s,n,Goto(s-${DIALSTATUS},1)
exten => s,n(voicemail),Voicemail(u${ARG1})
exten => s-NOANSWER,1,Voicemail(u${ARG1})
exten => s-NOANSWER,2,Congestion
exten => s-BUSY,1,Voicemail(b${ARG1})
exten => s-BUSY,2,Congestion
exten => s-CHANUNAVAIL,1,Voicemail(u${ARG1})
exten => s-CHANUNAVAIL,2,Congestion
exten => s-CONGESTION,1,Voicemail(u${ARG1})
exten => s-CONGESTION,2,Congestion
exten => s-.,1,Goto(s-NOANSWER,1)
exten => a,1,VoicemailMain(${ARG1})                              →35
[outbound]                                                       →36
include => dids
include => extensions
exten => _089XXXX,1,AgentLogin(${EXTEN:3
exten => _089XXXX,2,Hangup
exten => _011X.,1,AppendCDRUserField(calltype=international)
exten => _011X.,2,macro(stddid,${EXTEN})
exten => _1NXXNXXXXXX.,1,AppendCDRUserField(calltype=longdistance)
exten => _1NXXNXXXXXX.,2,macro(stddid,${EXTEN:1})
exten => _1NXXNXXXXXX,1,AppendCDRUserField(calltype=longdistance)
exten => _1NXXNXXXXXX,2,macro(stddid,${EXTEN:1})
exten =>     _NXXXXXX.,1,AppendCDRUserField(calltype=local)
exten =>     _NXXXXXX.,2,macro(stddid,${HOMEAC}${EXTEN:})
exten =>    _NXXXXXX,1,AppendCDRUserField(calltype=local)
exten =>    _NXXXXXX,2,macro(stddid,${HOMEAC}${EXTEN})
exten =>        _1X11,1,AppendCDRUserField(calltype=service)
exten =>        _1X11,2,macro(stddid,${EXTEN})
exten =>         _X11,1,Goto(1${EXTEN},1)
```

```
exten =>         _*88X.,1,Set(MONFNAME="${CALLERID(number)}-${EXTEN}-
                 ${STRFTIME(,,yyyy-MM-dd-hh-mm-ss)}")
exten =>         _*88X.,2,Monitor(wav|${MONFNAME})
exten =>         _*88X.,3,Goto(outbound,${EXTEN:3},1)                →56
exten => 9998,1,VoicemailMain()                                      →57
exten => 9998,2,Hangup
exten => 9999,1,VoicemailMain(${mbox})
exten => 9999,2,Hangup                                               →60
exten => _9[469]11,1,Dial(${TRUNK}/${EXTEN:1})                       →61
[dids]                                                               →62
exten => 2565551212,1,macro(stdexten,0001,SIP/0001)
exten => 2565551212,2,Hangup                                         →64
[extensions]                                                         →65
exten => _XXXX,1,macro(stdexten,${EXTEN}@default,SIP/${EXTEN})
exten => _XXXX,2,Hangup
exten => 5555,1,Set(MUSICCLASS()=customMOH)
exten => 5555,2,macro(stdexten,${EXTEN},SIP/${EXTEN})
exten => 5555,3,Hangup                                               →70
exten => 0,1,Queue(operator|t|||60)                                  →71
exten => 0,2,Queue(operator_overflow|t|||60)
exten => 0,3,Goto(0001,1)                                            →73
exten => 7000,1,MeetMe(${EXTEN})                                     →74
exten => 7001,1,MeetMe(${EXTEN},cMrx)                                →75
[default]                                                            →76
include => dids
include => extensions
exten => s,1,Answer()
exten => s,2,Wait(1)
exten => s,3,Background(custom/intro)
exten => s,4,WaitExten()
exten =>  1,1,Goto(queue-sales,s,1)
exten =>  8,1,Directory(default|f)
exten =>  9,1,Directory(default)                                     →85
exten => **9000,1,Authenticate(123456)                               →86
exten => **9000,2,Goto(recording-menu-main,s,1)
exten => t,1,Goto(default,s,2)
exten => i,1,Goto(default,s,2)                                       →89
[queue-sales]                                                        →90
exten => s,1,Background(custom/qsales-intro)
exten => s,2,Queue(sales)
exten => s,3,Goto(default,s,2)
exten => _X,1,Voicemail(u05001)
exten => t,1,Goto(default,s,2)
exten => i,1,Goto(default,s,2)                                       →96
[recording-menu-main]                                                →97
exten => s,1,Background(custom/recording-menu)
exten => s,2,WaitExten()
exten => 1,1,Record(custom/intro.gsm)
exten => 1,2,Goto(s,1)
exten => 4,1,Record(custom/qsales-intro.gsm)
exten => 4,2,Goto(s,1)
exten => 9,1,Record(custom/recording-menu.gsm)
exten => 9,2,Goto(s,1)
exten => *,1,Goto(default,s,2)
exten => t,1,Goto(s,1)
exten => i,1,Goto(s,1)                                               →108
```

We've removed all the spaces in between the contexts and their individual code, as well as the spaces between the end of a section of code and the start of a new context. You can add spaces in your dialplan to make it easier to read.

The dialplan consists of the following nine contexts:

→1–3   **Globals:** This context establishes information applicable to all contexts.

→4–17   **Macro-stdid:** Every Direct Inward Dial (DID) line used by the Asterisk requires routing, just like any extension. Instead of manually writing the code for each DID line, you can build a macro to reduce time and effort.

→18–35   **Macro-stdexten:** This standard extension macro allows you to eliminate the redundancy of writing and rewriting the code for every extension. This macro handles the decision making for any call to a standard extension that receives a no answer or a busy signal.

→36–61   **Outbound:** This context holds the routing plans that allow you to complete calls to the outside world.

→62–64   **DIDs:** This context lists all DID numbers considered local on your system in this context.

→65–75   **Extensions:** This context defines all the internal extensions associated with this Asterisk (we refer to it as the [internal] context in Part II). The sample dialplan identifies all 4-digit extensions as valid local extensions on an SIP device with voice mail. It also establishes the extension 0 to be forwarded to extension 0001.

→76–89   **Default:** The [default] context is also called the [inbound] context (see Part II). It is the boilerplate context that receives all inbound calls. The dialplan uses [default] to check all incoming calls to determine whether the number dialed is an extension or a DID number as well as to establish the routing of calls to invalid extensions.

→90–96   **Queue-sales:** This context handles the extraneous features outside the normal functions within the queue application. Tasks such as processing a digit-press from inside the queue or playing an introduction message prior to entering the queue are typically in this part of the dialplan.

→97–108   **Recording-menu-main:** The final context in the sample dialplan is boilerplate for the specific menu recordings possible after you enter the queue.

We discuss this dialplan in detailed in the rest of this appendix.

Going Global

Your dialplan should identify both the local area code for the Asterisk and the default device for all outbound calls. Depending on the depth and breadth of service provided by your Asterisk, you may be able to add more to this context, especially if you're using Asterisk as the office phone system for your company.

The [global] context includes only three lines of code:

```
[globals]                                                      →1
HOMEAC=256
TRUNK=SIP/111.222.333.444/                                     →3
```

The last line of code indicates the IP address for the VoIP device you use as the default trunk group for your outbound calls.

Maximizing DID Programming

We can't say enough good things about macros. Companies that provide VoIP phone service may have thousands, or tens of thousands, of DID numbers being handled by Asterisk. It is a blatant waste of time to write the 12 lines of code for each of your DID numbers. The macro context takes care of everything much more efficiently:

```
[macro-stddid]                                                 →4

exten => s,1,Dial(${TRUNK}${ARG1})
exten => s,2,Goto(s-${DIALSTATUS},1)
exten => s-NOANSWER,1,Congestion
exten => s-NOANSWER,2,Hangup
exten => s-BUSY,1,Busy
exten => s-BUSY,2,Hangup
exten => s-CHANUNAVAIL,1,Congestion
exten => s-CHANUNAVAIL,2,Hangup
exten => s-CONGESTION,1,Congestion
exten => s-CONGESTION,2,Hangup
exten => s-ANSWER,1,Hangup
exten => _s-.,1,Congestion
exten => _s-.,2,Hangup                                         →17
```

The initial `${TRUNK}` argument links the dialplan to the value of the trunk group, which sends the call to this macro. The `${ARG1}` portion of the argument is another expedient that links the call to the extension and voice-mail context of the DID number.

The `${DIALSTATUS}` argument for the `Goto` application establishes the logic for the handling of the call. If the status of the extension results in NOANSWER, BUSY, CHANUNAVAIL, CONGESTION, or ANSWER, the dialplan proceeds to the correct treatment based on that status.

Setting Up a Macro for Voice Mail

The standard extension (`stdexten`) macro handles routing to voice mail cleanly and efficiently. The `${ARG2}` argument in the following macro identifies the device(s) to ring. `${ARG1}` represents the extension dialed and voice-mail context just like in the DID macro.

The macro looks like this:

```
[macro-stdexten]                                               →18
exten => s,1(chanavail),ChanIsAvail(${ARG2})
exten => s,n,Goto(havechannel)
exten => s,chanavail+101,Goto(voicemail)
exten => s,n(havechannel),Dial(${ARG2},20)
exten => s,n,NoOp(${DIALSTATUS})
exten => s,n,Goto(s-${DIALSTATUS},1)

exten => s,n(voicemail),Voicemail(u${ARG1})

exten => s-NOANSWER,1,Voicemail(u${ARG1})
exten => s-NOANSWER,2,Congestion

exten => s-BUSY,1,Voicemail(b${ARG1})
exten => s-BUSY,2,Congestion

exten => s-CHANUNAVAIL,1,Voicemail(u${ARG1})
exten => s-CHANUNAVAIL,2,Congestion

exten => s-CONGESTION,1,Voicemail(u${ARG1})
exten => s-CONGESTION,2,Congestion

exten => s-.,1,Goto(s-NOANSWER,1)

exten => a,1,VoicemailMain(${ARG1})                            →35
```

The standard extension macro uses the Asterisk default of establishing a voice-mail priority 101 digits above the existing priority for the specific extension dialed. For example, this allows you to dial extension 133 and have the call routed through priorities 1 and 2, and then jump to priority 103 (priority 2 +101) if the call isn't answered. The fourth line in the code dials the voice-mail extension if the call is unanswered in 20 seconds, as represented by the following:

```
Dial (${ARG},20)
```

Dialing the Outside World

The first two line items of the `[outbound]` context show the `[dids]` and `[extensions]` contexts. Asterisk can make these connections itself without sending the call out to a carrier, where you'll have to pay a per-minute charge:

```
[outbound]                                                              →36
include => dids
include => extensions


exten => _089XXXX,1,AgentLogin(${EXTEN:3})
exten => _089XXXX,2,Hangup

exten => _011X.,1,AppendCDRUserField(calltype=international)
exten => _011X.,2,macro(stddid,${EXTEN})
exten => _1NXXNXXXXXX,1,AppendCDRUserField(calltype=longdistance)
exten => _1NXXNXXXXX.,2,macro(stddid,${EXTEN:1})
exten => _1NXXNXXXXXX,1,AppendCDRUserField(calltype=longdistance)
exten => _1NXXNXXXXX,2,macro(stddid,${EXTEN:1})
exten =>    _NXXXXXX.,1,AppendCDRUserField(calltype=local)
exten =>    _NXXXXXX.,2,macro(stddid,${HOMEAC}${EXTEN:})
exten =>    _NXXXXXX,1,AppendCDRUserField(calltype=local)
exten =>    _NXXXXXX,2,macro(stddid,${HOMEAC}${EXTEN})
exten =>     _1X11,1,AppendCDRUserField(calltype=service)
exten =>     _1X11,2,macro(stddid,${EXTEN})
exten =>      _X11,1,Goto(1${EXTEN},1)

exten =>       _*88X.,1,Set(MONFNAME="${CALLERID(number)}-${EXTEN}-
               ${STRFTIME(,,yyyy-MM-dd-hh-mm-ss)}")
exten =>       _*88X.,2,Monitor(wav|${MONFNAME})
exten =>       _*88X.,3,Goto(outbound,${EXTEN:3},1)                     →56
```

Line 36 provides a login extension for the employees (*agents*) that handle calls from the `[queue-sales]` context, which is farther down in the dialplan. Line 37 enables the agents to log off the queue.

The bulk of this context assigns a call type to the call based on the dialing sequence and checks to see whether the number being dialed exists in either the `[dids]` or `[extensions]` context. Pattern matching appends the call type into the CDR file as follows:

- Calls placed with 011 are branded as international.
- 10-digit dialed numbers with a leading 1 are listed as long distance.
- 7-digit numbers are tagged as local, with the number being translated to a full 10-digit number by adding your local area code (HOMEAC) to the 7-digit dialed number.
- Service calls are identified with the pattern matching of 1X11, which catches any call to 411, 611, or 911.

The following code sets the monitor filename (MONFNAME) in the call monitor section of the code:

```
_*88X.,1,Set(MONFNAME="${CALLERID(number)}-${EXTEN}-
          ${STRFTIME(,,yyyy-MM-dd-hh-mm-ss)}")
```

The second priority of that sequence begins the monitoring of the call, and the final line of code strips the leading 88 digits from the extension dialed. This grouping allows you to dial any extension with a leading 88 to initiate a call monitor.

The next portion of the `[outbound]` context gives you convenient access to your voice mail. Dialing extension 9998 allows you to input your voice-mail extension (in case you want to pick up someone else's voice mail while he is on vacation). Extension 9999 is a direct connection into the voice mailbox for your extension, as long as you've set up the `setvar=mbox=yourextension` line in your device definition. The final element of this context is the line of code directing the calls dialed to 911, 611, and 411.

This section of code is as follows:

```
exten => 9998,1,VoicemailMain()                           →57
exten => 9998,2,Hangup

exten => 9999,1,VoicemailMain(${mbox})
exten => 9999,2,Hangup                                    →60
exten => _9[469]11,1,Dial(${TRUNK}/${EXTEN:1})            →61
```

Integrating Direct Inward Dial Numbers

VoIP service isn't commonly bidirectional at this time. The legal regulations for outbound VoIP service prevent carriers from releasing outbound service along with inbound service. This isn't a large problem because VoIP is dynamic. You can establish your phone number for inbound calling, and all outbound service can be provided based on the IP address from which you originate your call. In this case, your phone number may be attributed to the outbound call, but any number could be linked to your IP address and input in the space.

The bottom line is that if you use your Asterisk to handle VoIP calls, you'll probably collect a lot of DID numbers. You can differentiate them from your standard extensions by placing them in their own context, but treat them much like any other extension:

```
[dids]                                                          →62
exten => 2565551212,1,macro(stdexten,0001,SIP/0001)
exten => 2565551212,2,Hangup                                    →64
```

Routing to Extensions

Your extensions context probably begins predictably. You establish a line of code to accept any 4-digit dialed extension and send it to voice mail if there's no answer. You can then set the music on hold and the foundation has been set:

```
[extensions]                                                    →65
exten => _XXXX,1,macro(stdexten,${EXTEN}@default,SIP/${EXTEN})
exten => _XXXX,2,Hangup

exten => 5555,1,Set(MUSICCLASS()=customMOH)
exten => 5555,2,macro(stdexten,${EXTEN},SIP/${EXTEN})
exten => 5555,3,Hangup                                          →70
```

You can add some spice to your dialplan by building an operator queue with overflow:

```
exten => 0,1,Queue(operator|t|||60)                             →71
exten => 0,2,Queue(operator_overflow|t|||60)
exten => 0,3,Goto(0001,1)                                       →73
```

And then you can finish off your extensions context by establishing a conference room. The second line of code simply enables the conferences to be recorded:

```
exten => 7000,1,MeetMe(${EXTEN})                                →74
exten => 7001,1,MeetMe(${EXTEN},cMrx)                           →75
```

Providing a Default Context

Your [default] context acts as a catchall for calls that aren't dialed to a predefined extension. You begin by including the [dids] and [extensions] contexts along with a standard progression of priorities for your default extension:

```
[default]                                                       →76
include => dids
include => extensions

exten => s,1,Answer()
exten => s,2,Wait(1)
exten => s,3,Background(custom/intro)
exten => s,4,WaitExten()

exten =>  1,1,Goto(queue-sales,s,1)
exten =>  8,1,Directory(default|f)
exten =>  9,1,Directory(default)                                →85
```

Your incoming calls receive an intro recording that asks the caller to input the DID number or extension he wishes to reach. The dialplan provides the option to go to the sales queue or to use the company directory by first or last name.

You can add something more to help you with your intro greeting. The **9000 extension listed in the continuation of the [extensions] context allows you to record your intro message:

```
exten => **9000,1,Authenticate(123456)                          →86
exten => **9000,2,Goto(recording-menu-main,s,1)

exten => t,1,Goto(default,s,2)
exten => i,1,Goto(default,s,2)                                  →89
```

We show you how to enable this feature in the section "Recording a Custom Menu," later in this appendix.

Queuing Up Company Sales

This context uses a standard queue application to channel incoming calls to a pool of available sales agents. It plays the custom menu prompts called `qsales-intro` and walks the calls through a normal progression:

```
[queue-sales]                                        →90
exten => s,1,Background(custom/qsales-intro)
exten => s,2,Queue(sales)
exten => s,3,Goto(default,s,2)

exten => _X,1,Voicemail(u05001)

exten => t,1,Goto(default,s,2)
exten => i,1,Goto(default,s,2)                       →95
```

Recording a Custom Menu

Every business model is different. Your company may be a think tank or financial institution that doesn't have a traditional distribution and sales environment. Your company might differentiate itself from other companies in the industry by its creative flair or staunch professionalism. Regardless of the situation, use a prompting menu that reflects the philosophy of your business. This code allows you to record your own greeting, which you can then apply to any queue or section of your dialplan:

```
[recording-menu-main]                                →97
exten => s,1,Background(custom/recording-menu)
exten => s,2,WaitExten()

exten => 1,1,Record(custom/intro.gsm)
exten => 1,2,Goto(s,1)

exten => 4,1,Record(custom/qsales-intro.gsm)
exten => 4,2,Goto(s,1)

exten => 9,1,Record(custom/recording-menu.gsm)
exten => 9,2,Goto(s,1)

exten => *,1,Goto(default,s,2)

exten => t,1,Goto(s,1)
exten => i,1,Goto(s,1)                               →108
```

Appendix B

VoIP Basics

The greatest attribute of Asterisk is its flexibility. As much as we'd like to say that flexibility is the magic of Asterisk code, it's only part of the reason for the Asterisk's versatility. A great deal of the Asterisk's power lies in its ability to integrate with the greatest telephony breakthrough of our time: Voice over Internet Protocol, more commonly known as *VoIP*.

VoIP is the most profound change to hit telecom since the advent of the dial tone. It integrates the structure and methodology usually attributed to data transfer to the massive requirements of completing voice calls. This chapter covers the basics of VoIP as it pertains to Asterisk.

Finding Out How VoIP Works

VoIP, at its most basic level, is nothing more than the transmission of a voice phone call using the Internet Protocol (IP). It is a bit more complex than just sending an e-mail or an instant message, but with the correct software such as Asterisk, VoIP isn't much more complex.

Every phone call (whether VoIP or non-VoIP), has the following two essential elements (see Figure B-1):

✔ **The voice portion:** This is the part of the call in which your words are transmitted. Every nuance of your speech and every pause, laugh, or inflection of your voice is converted and sent in this part of the call.

The voice portion is continually transmitted and can be transmitted from any port from 10000 to 20000. We randomly choose port 10950 and 15270 for Figure B-1. Also note that the voice portion is being sent to the Asterisk from a media server, and not the Edge Proxy Server (EPS) from which you receive the overhead of the call. VoIP technology allows two completely different pieces of hardware to handle each aspect of the call. The EPS transmits the overhead for the call from the edge of your carrier's network to your Asterisk. Their media server is tasked with handling the transmission and receipt of the voice portion of the call.

This is one of the fundamental differences between traditional (non-VoIP) phone calls and VoIP phone calls. The overhead and the voice portion of the call don't have to be transmitted together from the same device.

✔ **The overhead:** This is where the non-voice transmission takes place. The transmission of caller ID, the connect signal that begins the billing of the call, and the disconnect signal that identifies the call as being complete are all transmitted in this section of the call.

The paths taken by the overhead are represented in dotted lines because little overhead is required after the call is established. The industry-standard port 5060 handles the overhead.

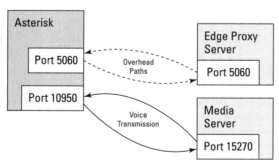

Figure B-1:
A VoIP schematic showing the overhead and voice paths.

The majority of bandwidth needed to place a call is consumed in transmitting the voice portion of the call. The overhead requires about 8 Kbps of bandwidth versus the 56 or 64 Kbps of bandwidth that the voice portion needs to transmit.

Using SIP for VoIP

Session Initiation Protocol (SIP) is the preferred method of handling VoIP calls with Asterisk. Most VoIP carriers prefer you to connect with SIP instead of the older protocol called H.323. There are a couple reasons for this:

✔ SIP is structurally faster than H.323. It uses less time setting up calls, requiring only one invite message, versus the eight messages for H.323.

✔ SIP is an easy-to-read text-based protocol. H.323 is based on a binary software code that is difficult to read and comprehend. The text-based nature of SIP enables it to be easily dissected and analyzed, directly as it is transmitted from your Asterisk.

SIP is broken into the following two protocols:

✔ **Session Description Protocol (SDP):** This is the part of SIP that describes the call initiation and invitation (involving the negotiation with the remote server of both: the port from which to send and receive information, as well as acceptance of the call being sent). It is also the part of SIP used to reinvite calls.

✔ **Real-Time Transport Protocol (RTP):** This is a transport protocol that tells the VoIP equipment how to package the data and convert it on the far end of the call. It transmits the voice part. It is a separate aspect of SIP and can even function separately from the SDP portion of the call.

Building a standard VoIP call

The most basic VoIP call involves a remote VoIP phone, your Asterisk server, and the EPS of your VoIP carrier routing your call to the VoIP (or even a non-VoIP) phone at the far end. Figure B-2 shows the elements of a basic VoIP call.

Figure B-2:
A standard
VoIP call.

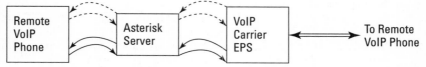

This standard call flow is useful, but not very efficient. The amount of bandwidth necessary to process the RTP stream between your carrier's EPS and the remote VoIP phone on your end can quickly consume all your available IP bandwidth. This basic scenario works fine when servicing only a few remote VoIP phones, but as your remote VoIP connections increase, you'll quickly run out of bandwidth. Luckily, there is an easy solution available with VoIP; it's called *re-inviting*.

Maximizing bandwidth by reinviting calls

VoIP calls are connected when the device that originates the call sends an INVITE message (or *invitation*) to the device that receives the call. After the device acknowledges the connection, the call isn't locked in to that end device. Figure B-3 shows the progression of a call from the time your Asterisk server receives the call to the point where it is forwarded to a remote VoIP phone (a *reinvite*).

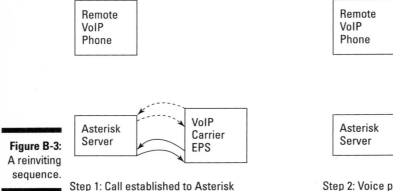

Figure B-3:
A reinviting
sequence.

Step 1: Call established to Asterisk   Step 2: Voice portion of call reinvited

The diagram on the left identifies an inbound VoIP call that's established between your VoIP carrier and your Asterisk server. The dotted lines represent the path taken by the overhead, and the solid lines represent RTP handling the voice portion of the call. The call has the RTP (voice) and the SDP (overhead) spanning the two devices.

In between the left diagram and the right diagram, the dialplan in your Asterisk is taking action. The phone number dialed on the incoming call is identified as the remote device and Asterisk sends a request to re-invite the RTP of the call to the remote VoIP phone. Your carrier's EPS acknowledges the request for the reinvite and redirects the RTP to the new end point.

The diagram on the right shows the overhead portion of the call still spanning your VoIP carrier to your Asterisk service, but the RTP (voice) portion is now sent to a remote VoIP phone that could be anywhere in the world. The beauty of VoIP is the fact that the voice portion of the call can be sent to a remote location independent of the devices that are handling the overhead.

The two main benefits of reinviting phone calls are as follows:

✔ **You eliminate a series of elements that could cause audio problems.**
Every device that handles the RTP of a VoIP call has the potential of
adding *latency* (or delays), echo, or static or dropping packets. VoIP is a
sensitive technology when it comes to latency. The more hardware that
handles a call, the more devices the call negotiates to complete and the
more latency the call incurs. After about 200 milliseconds of latency, the
quality of your call begins to deteriorate. Every piece of hardware you
remove from the transmission is another bit of latency that you avoid.

On a strictly troubleshooting level, every piece of hardware that handles
a call is also a potential source of static and echo that you must methodically clear if you encounter quality issues. We suggest eliminating
unnecessary hardware in the call by allowing the reinvite process.

✔ **You save bandwidth.** Figure B-4 demonstrates the bandwidth you save
when reinviting a call directly to the remote VoIP device.

Figure B-4:
A bandwidth
comparison
of reinvited
and non-
reinvited
calls.

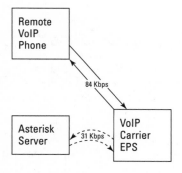

Reinvited call – Asterisk bandwidth – 31 Kbps

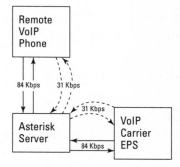

Non-reinvited call – approximately 230 Kbps

The call on the left is a reinvited VoIP call. The Asterisk server forwards
the RTP of the call directly to the remote VoIP phone, and the server is
now receiving periodic overhead information that consumes a few Kbps
of bandwidth.

The call on the right is an VoIP call that isn't reinvited. The Asterisk server
is managing a separate pair of RTP streams for the second leg of the call
and an additional overhead connection, consuming an additional 199 Kbps
in the process.

Some carriers aren't fond of reinviting calls. Every time a call is invited or reinvited, the carrier's EPS is engaged to establish the new RTP stream. Although this process takes only a few milliseconds, it does occupy that part of the EPS, preventing other calls from using the same section of the interface to establish a call. At this time, VoIP isn't widespread enough that any carrier's EPS is overloaded. The additional three or four packets sent by your hardware to initiate the reinvite are just a drop in the bucket compared to the millions of calls per day that the carrier's EPS processes.

The independent nature of the RTP stream has a dark side. *Rogue RTP* is a condition where an SIP device invites the RTP stream to the incorrect IP address. The result of such an uncalculated invite is an RTP stream flying across the ether, crashing into an unsuspecting server somewhere in the world. The barrage of RTP data can either severely slow the speed of the receiving server or potentially crash the server. The process of inviting and reinviting calls is solid with Asterisk; it's unlikely that you can cause rogue RTP. The RTP is only sent to a remote server after a handful of codes have passed between the two servers, validating both parties.

Evaluating your compression options

To minimize the bandwidth that a call needs, you need to *compress* (shrink) your calls. For compression, you need to choose a codec. A *codec* (short for *co*der-*dec*oder) converts your voice call to or from a digital code that can be sent over an Internet circuit.

You have several choices in codecs; all of them have benefits and detriments. When choosing the correct codec for your application, remember the following three elements:

- ✔ **Compression ratio:** Compression is the financial silver lining of VoIP. The greater the compression that's possible on a call, the more calls you can place on a circuit. Traditional telephony only allows 24 active calls on a single DS-1 circuit before you must buy another circuit from your carrier. That additional circuit involves an installation fee, monthly fee, and the need to purchase more hardware to connect to the circuit. By using one VoIP DS-1 for the same quantity of active calls as two or more traditional DS-1s, you prevent any additional expense.

- ✔ **Call quality:** Some compression techniques reduce call quality. These two elements are generally inversely proportional. The goal is always to achieve the greatest compression with the best call quality.

- ✔ **Packetization delay:** Whenever a call must be converted from VoIP to analog or vice versa, the process takes some time. In the realm of your daily life, the 50 milliseconds that a codec takes to compress and packetize a voice call are inconsequential; in the VoIP world, the time delay can be an issue.

Two codecs stand above the rest: The uncompressed codec G.711 and the 8:1 compression codec G.729 have been adopted as the industry standards. Other codecs are available and supported by VoIP carriers, but the G.711 and G.729 are by far the most common codecs.

All SIP hardware lists the codecs they accept in descending preferential order. The hardware that connects to your Asterisk also has a list of the codecs that it supports in descending order. Because the benefit of VoIP is its ability to maximize bandwidth, your SIP hardware should list G.729 as the first codec of choice, with G.711 as the backup. This configuration allows your hardware to initially negotiate for the best bandwidth with G.729, but fall back to the stable, uncompressed G.711 codec. Because all carriers support G.711, always request the compressed codec before you settle for the uncompressed codec.

We go into more detail about how to choose the right codec in Chapter 9.

Optioning for touch tones

The lovely sounds you hear when you push the touch-tone buttons on your phone are more complex and amazing than you know. The telecom industry refers to touch tones as *Dual Tone Multi-Frequency (DTMF)* tones. This is because the sound generated by each button on your keypad is actually two different tones being sent at two different frequencies. The different frequencies allow phone systems to differentiate the touch tones from the sound of a human voice.

VoIP has three ways it can process DTMF tones. In order of preference, they are as follows:

- **RFC 2833:** This method lists the tones as DTMF events in the overhead of the RTP stream. The tones themselves aren't sent; a message in the overhead of the RTP states that a DTMF event is requested. (RTP isn't just the transmission of raw voice data. It does include a thin overhead to manage the voice portion of the call, and also potentially handles DTMF.) The receiving hardware, be it a VoIP phone on the LAN with your Asterisk or a VoIP phone on the other side of the world, accepts this information and plays a recorded file for the DTMF tone so that you hear what you expect.

- **SIP INFO:** This method is like RFC 2833 in that a DTMF event message is sent instead of an audible tone. The difference is that the SIP INFO method sends the DTMF event in the SDP overhead of the SIP instead of the RTP overhead.

- **In-band:** This method is straightforward. The DTMF tone is sent in the voice portion of the call, just as if a person were speaking.

In-band transmission is the least desirable because any audio distortion in the call also affects the DTMF tones and can cause problems. DTMF tones are complex, and any degradation in them could result in the tone being unrecognized. RFC 2833 and SIP INFO are contained in the overhead and are unaffected by the manipulation of the voice portion of the call.

Passing the InterOperability Test

VoIP technology has standards for everything, but every carrier and piece of hardware may use a slightly different interpretation of a portion of the standard. For example, your hardware may only accept a 10-digit phone number, but your carrier sends the information with "+1" preceding the number because that is the international standard.

E.164 is the internationally accepted standard for transmitting phone numbers with VoIP. It specifies a maximum of 15 digits, consisting of a country code, a city code (referred to as a *National Destination Code*), and the subscriber number.

Carriers understand that nuances exist between the configuration of their EPS and individual routers and servers of their customers. The unique business applications and features required by customers mean that minor adjustments probably need to be made. To eliminate a protracted period of opening trouble tickets for every variety of call and feature, the carrier often has an InterOperability test (or *InterOp test*) you must complete before you are released onto its network.

When you sign up for service, your carrier sends you an InterOp questionnaire form to assist in configuring its hardware to match your Asterisk. Fortunately, Asterisk isn't a custom-built, on-off software that's completely unknown to VoIP carriers. They all have customers with Asterisk systems established on their networks, and the software has a solid history of interoperability. You shouldn't encounter issues that can't easily be resolved. Most InterOp tests require, at a minimum, the following information from you:

- ✔ **Your Internet Protocol (IP):** This is SIP.

- ✔ **Your ISP:** Select a quality Internet Service Provider for your VoIP. An inexpensive ISP is acceptable if you have a dialup connection to surf the Web and send e-mails, but is unacceptable for VoIP transmissions. Many carriers don't require you to order IP connections to their network as long as you are using a quality carrier.

- ✔ **The location and IP address of your SIP hardware:** Your carrier must load your IP address into the correct server on its network to allow your IP address access. Even if the carrier gives you the IP address to its SIP hardware first, you're still blocked by its firewall until your carrier has your IP address established on its network.

✔ **The location and address of your media gateway:** You may have a separate device that handles the RTP of your calls that your carrier must also program into its network. You can expect one-way audio on your calls if you don't provide this information.

✔ **Other SDP and/or RTP hardware:** Many carriers are interested in the hardware on the back end of your Asterisk server. The environment in which your server resides (such as a busy LAN) can be a source of latency, jitter, or lost packets. The InterOp test is much simpler if you reveal the other variables that weigh on your LAN.

✔ **Codecs:** You must decide whether you want the G.711 and G.729 codecs, or whether you also need G.723 and G.726. If you're sending faxes, you also need T.38.

✔ **DTMF:** Your carrier needs to know your DTMF transmission method of choice: RFC 2833, SIP INFO, or in-band.

The InterOp tests can be tedious and complex or simple and easy; it depends on the carrier. Level 3 has the most rigorous and methodical InterOp test, but Qwest is much more relaxed. The intent of the process is to provide a controlled environment where you can test every codec, feature, and type of call you envision ever using and resolve potential issues.

A simple InterOp test should include a battery of calls for each codec being used, including the following:

✔ An inbound call where the call completes and the origination party hangs up

✔ An inbound call where the call completes and the receiving party hangs up

✔ An outbound call where the call completes and the origination party hangs up

✔ An outbound call where the call completes and the receiving party hangs up

✔ An outbound call where the origination party hangs up before the receiving party answers

✔ A call in which the origination party sends DTMF (touch tones)

✔ A call in which the receiving party sends DTMF (touch tones)

✔ An inbound fax transmission

✔ An outbound fax transmission

Faxes don't work on G.729, but be sure to test them on G.711 and T.38 if you plan to send or receive them.

Accepting or ignoring SIP-T

SIP-T is a feature of SIP that sends the industry-standard 2-digit code that identifies the type of phone line that is originating the call. SIP-T is only useful if you need to identify inbound calls that come from a prison, hospital, or pay phone, for example. SIP-T is an enhancement to SIP that supports SS7, a specialized protocol used within large telecom networks. At this time, Asterisk doesn't support the SS7 information, but a development version is in the works. Just say no to SIP-T for now.

VoIP Quality Issues

VoIP phone calls, just like other phone calls, have the potential for both quality and completion problems.

Completion problems are probably the result of a problem on the call beyond the point where it hits your VoIP carrier and enters the Public Switched Telephone Network (PSTN). The problem most likely doesn't reside within your network. You can confirm that the call is reaching your carrier by performing a packet capture on a failed call (see Chapter 8).

Call-quality issues, such as echo, static, or *clipping* (where you only hear bits and pieces of a conversation, like with a bad cellphone connection), can be the result of your LAN or your IP provider. These quality issues may be the direct result of configuration settings on your side. Echo, for example, can be caused by the transmit volume being set too high on your Asterisk. For the most part, the tedious quality issues like static and clipping are caused by the two biggest concerns in VoIP: latency and jitter.

We discuss VoIP call quality issues in depth in Chapter 12.

Moving from Non-VoIP Service to VoIP Service

Moving your local phone number from a non-VoIP carrier to a VoIP provider can take from eight days to eight weeks (or longer in some cases). You need to sign and date a Letter of Authorization/Agency (LOA) provided to you by your new carrier. The company must have the letter on file so that it can move your phone number from your old carrier (even if your old carrier doesn't require the letter) to its VoIP service. You may also have to supply a copy of your phone bill if the migration is challenged.

Every phone number doesn't have to be migrated from your old local carrier to a new one. We recommend only moving essential numbers that are printed on your business cards, paperwork, or are established with your customers. Most companies have one primary phone line that rings to their office with a handful of other lines to accept overflow if the first line is busy. Because nobody dials these numbers directly, it's easier to replace them than to contend with the challenge of migrating them.

Federal law prevents your current carrier from rejecting the request to migrate your phone number. However, in reality, orders are frequently rejected. Here are some of the reasons why orders are rejected:

- ✔ **A name mismatch occurs.** The name on the account with the carrier doesn't match the name listed on the order.

- ✔ **An address mismatch occurs.** The address on the account with the carrier doesn't match the address listed on the order.

- ✔ **All the data mismatch.** The address and name don't match the account with the carrier for the phone number. This indicates you should recheck the phone number you're trying to migrate. You may have mistyped the phone number on the LOA and it might belong to a completely different company.

- ✔ **Features can be lost if the number migrates.** Many phone lines have voice mail, caller ID, conference calling, call waiting, or other features that must be removed before the line can be migrated. None of the existing features migrate with a phone number, and the new carrier must build them in again. Some carriers reject a migration request because of the features present on the line. If your number is rejected for having features, call your old local carrier and place an order removing all the features. After the order completes, resubmit the migration and it should complete.

- ✔ **Your number is part of a Centrex group.** Centrex groups are special bundled packages of phone lines. You must have your number separated from the Centrex group before submitting it for migration.

- ✔ **Your number is associated with digital subscriber line (DSL) service.** You must separate your DSL service from any phone number before submitting your number for migration.

- ✔ **Your number is associated with other phone numbers.** Similar to Centrex lines, phone numbers are frequently established in groups. You must cancel all other numbers in the group or migrate them.

- ✔ **Your number is a virtual number.** Some phone numbers only exist to provide a service like distinctive ringing. These numbers are not portable.

Understanding resold accounts

The world of telecom is increasingly complex. A local phone carrier such as Bell South may own a phone number, but the person who is using it may have ordered the phone line from a third-party company who resells Bell South. This is where local number portability becomes complex.

For example, suppose that you have phone service from a company called TelecomReseller that uses the Ameritech network. What is the right name to place on the LOA that you supply to your new carrier? The request to have your phone number released is sent to the true owner of the phone number — in this case Ameritech (which believes that TelecomReseller owns the number, not you). Ameritech promptly rejects the order because your name and address don't match the name and address in its records (TelecomReseller). To resolve the conflict, you need to supply a copy of your most recent invoice for the phone number. Even then, it could get more complex and take more time.

To make matters worse, the company you are moving to may also be a reseller of another phone carrier, further muddying the mix. Just be aware of the fact that several layers of companies can be involved in the ownership of your phone number. Every company added to the equation means another possible source of confusion or rejection. Be ready to supply documents and make conference calls if necessary.

This seems like a wonderful laundry list of ways your carrier can prevent you from moving your service to another carrier, but they are all essential checks and balances. The system also prevents other people from accidentally migrating your phone number because they miskeyed the phone number or were unaware of the fact that moving the line would disconnect your DSL service that may be the lifeblood of your business.

You are given a Firm Order Commitment (FOC) date when the number is set to migrate. The FOC date identifies the day on which the number is given to your new carrier. At 9:00 a.m. on the given day, like magic, your number is active with the new carrier. If the releasing and gaining carriers have a friendly relationship, it may only take eight days from the time you requested it to be moved for the number to become active. In this case, rejoice at your good fortune. If the change isn't that quick, prepare yourself for what could be a few weeks of submitting documents and negotiating the release.

Appendix C

Understanding Basic Linux

. . . .

. . . .

*J*ust like your Microsoft Word runs under Windows XP, the Asterisk software you're using runs on a server that uses Linux as the operating software. Understanding how Linux works is important because you set up Asterisk in Linux. The very nature of the open-source architecture of Asterisk lends itself to modification and manipulation, even after the initial configuration. If you don't feel comfortable working in Linux, we help you get your bearings. Linux has a wealth of commands and features available to it.

Linux is wonderful, powerful, and complex software. The information provided in this appendix is only the thinnest veneer of education that allows you to stumble through an initial setup of Asterisk. Your level of frustration with the Asterisk software is going to be inversely proportional to your knowledge of Linux and general programming. If you are responsible for all Asterisk development and management, you must expand your knowledge of Linux. You might be able to lean on some of your friends that know Linux for a while, but in the end, the programming rests on your shoulders. Take some classes, do some research, and find out as much as you can. Getting to know one small function or macro could take hours off your workload. As trite as it sounds, in the world of complex software, you really can work smarter, not harder.

Defining Linux

Linux is one of the most popular operating systems for server applications. The software has hundreds of different distributions (*distros*) that consist of the same base elements, with each distribution having its own specific software twist.

Linux is a compilation of the following three elements:

- **Operating system kernel:** The *kernel* is the brain that enables all the software on your system to interact with your hardware. The display on the monitor of the words you type and the tracking and clicking of your mouse are all possible due to interaction with the kernel. The driver for your mouse, as well as drivers for all the other hardware connected to your computer, are loaded with the kernel, which allows them to function.

- **Application software:** This element is what people generally see as software. Application software isn't required to run your computer or additional software. Video games are the most obvious example of application software. They aren't necessary for the computer to work or for other software programs to function. They are the end recipient of the kernel, hardware drivers, and supporting software.

- **System software:** The system software supplements the kernel, performing everyday tasks and assisting the other software on the system to function. This software is functionally positioned between the kernel and the application software. Software used for logging data is a prime example of system software. It writes data to a log file for application software. It doesn't interact with the hardware elements like the kernel, but it does bridge the application software–to–kernel functionality.

Most people refer to the entire package of the previous three elements when talking about Linux.

The commands and elements we discuss are common to all distributions. If an in-house staff maintains your Asterisk, ensure that they know about your specific distro. The speed at which your staff can make changes and resolve problems may depend on their familiarity with the software.

Navigating in Linux

Moving around in the Linux environment can take one of the following two forms:

✔ **Console:** This option looks like the old DOS interface or a command prompt in Windows. It is also referred to as the command-line interface (CLI).

✔ **Graphical User Interface (GUI) mode:** The specific nuances of the GUI interface available vary based on the window manager and the theme. *Window managers,* such as Gnome or KDE, interact with the Linux software, while the *theme* is geared toward the visual representation of the GUI. Looking at it from a business standpoint, the window manager is the operations department that makes everything work, while the theme is the marketing department that puts the pretty face on for the customer.

Don't use the X-Windows GUI for Linux on your server running Asterisk. It may seem like a lot of fun, but it is a huge drain on resources.

Logging in to Linux in Console mode deposits you at either your *home directory* if you signed in to a user account or to the `/root/` directory when signing in as a *root* user.

The home directory is generally found at the following location:

```
/home/username/
```

Login and passwords are required for all Linux logins in both GUI and Console mode. The root directory is a *super user* account with access to all aspects of the Linux system. The exception is if you're running an SE version of Linux. Linux SE uses even more advanced security implementations that can restrict root user access. Be careful when working with an SE version of Linux because you could secure the system so well that you prevent root access to vital aspects of the kernel. After you lock yourself out of the root access on the system, your only option is to rebuild the entire system.

After the Linux CLI is open, use the following commands to navigate to other directories:

✔ `ls`: Lists the files and subdirectories within the current directory.

✔ `ls -l`: Lists the same files and directories as the `ls` command, but in a long format that lists the permissions on the files.

✔ `ls -a`: Shows not only the regular files but also the hidden files in the directory. The hidden files may not necessarily be hidden, but their names may start with a period (`.`) and don't appear when you run the `ls` command.

✔ `ls | more`: Presents all the files and subdirectories within the current directory, one page at a time.

✔ ls | less: Displays the information a page at a time, scrolling backward.

✔ ls -R | more: Use this command when you are looking for a file or directory and you don't know where you saved it. The ls -R portion of the command lists not only the files and subdirectories in the directory you are in but also the contents of each subdirectory. The list invariably includes over a page of data, so the | more addition to the command allows you to receive the data in a friendly manner.

✔ cd: Allows you to move from one directory to another within Linux.

The pipe (|) on your keyboard (above the Enter key) instructs Linux to redirect the output of the first command to the parameters of the second command. The ls | more command is understood by Linux to mean "List all files and subdirectories and export them to the screen one page at a time based on the more command." Pressing the spacebar displays another page of data until no more data is available.

Introducing Permissions

The long-file format presented by using the ls -l command reveals more of the complexity of Linux. It identifies the owner of each file and directory and shows the type of access available to each group.

The following code lists two directories, named ProgramOne and WorkingFolder, and a file named GeneralUse:

```
drwxr--r-- 3 Brady stargat 3645 Jul 1 11:43 ProgramOne
drwxr-xr-x 2 Brady stargat 2741 Sep 4 17:23 WorkingFolder
-rwxr-xr-x 1 Brady stargat 355 Mar 23 12:15 Generaluse
```

The tightly packed columns to the left indicate the permissions allowable to three groups. The available permissions are as follows:

✔ r: This indicates the group or person who has the authority to read the contents of the file or directory.

✔ w: This indicates the group or person who has the authority to write new information to or edit a document.

✔ x: This represents the ability to execute the file or directory, allowing the person or group to launch the program.

✔ -: This represents the inability to read, write, or execute a file or document, depending on where the - is located. The sequencing is always the same, so a permission of r-x indicates the ability to read and execute a program, but not to edit or write to the program.

Linux identifies the following three entities, each having their own permissions:

- ✔ **Owner:** The first grouping is owner permissions. The *owner* of a file or directory is the individual who created it. For the `ProgramOne` directory, the owner permissions are `rwx`. This owner can read, write, and execute.

- ✔ **Group:** The *group* is the specific group to which the owner of the file belongs. The owner of the files is Brady, and he belongs to the group named `stargat`. For the `ProgramOne` directory, the group permissions are `r--`. Anyone belonging to the `stargat` group can read but not write or execute.

- ✔ **World:** This is a generic term for anyone on the Linux system who isn't either the owner or a member of the owner's elite group. For the `ProgramOne` directory, the world permissions are `r--`. Anyone who isn't the owner or part of the group can read but not write or execute.

Root access provides full access to read all files. It doesn't guarantee that you have write or execute permissions. Root access can be used to leverage the ability to write a file, but you won't be able to execute a file unless you are specifically granted permission to do so.

Navigating Directories

Linux has a methodical directory system that makes file placement predictable. The *root* directory is the main directory for the system and is represented by a single backslash. The root directory contains all the other directories; the concept of *root access* is important to understand when working with Linux.

Signing in as a root user delivers you to the root directory. If you're working in Linux and need to return to the root directory, use the following command:

```
cd/
```

The root directory is rather boring, consisting of nothing more than a single (forward) slash (/). You can dress up your Linux prompt by adding your host name and a character to identify the end of the prompt, as shown in the following example:

```
Venus/ #
```

This is a bit more dynamic than a boring single slash. In this case, our domain name is Venus and we're using the # sign to signify the end of our prompt.

Knowing root from root home

Don't confuse the root directory and the root home directory. The root directory prompt looks like this:

```
/
```

The root home directory looks like this:

```
/root/
```

In spite of their similar names and appearance, they are quite different. The / root directory is the base directory that contains the other directories in Linux. It is the alpha and omega of the Linux world, the undisputable top of the pyramid from which everything else lines up underneath. The /root/ directory is the typical home directory for the root user accounts. This directory has the specifics on all the individual root users but isn't the tip of the iceberg for Linux directories.

Root access is important because the Asterisk software you're running is installed as root. After you install the Asterisk software, you don't need root access for the day-to-day operations. The only time you need root access after installation is when you need to execute packet captures.

If you do need to do something in root but are leery of using it, you can borrow root access through the following methods:

- ✔ **The** sudo **command:** This command allows you to gain temporary root access for a specific command or process. When you complete your work, you automatically return to standard user access.

- ✔ **The right user group:** You can create user groups that allow special permissions for specific tasks. You can set up just enough permissions to accomplish the task and not expose Linux to accidental tampering.

The least secure option is requiring individuals to log in as root, accomplish the task at hand, and then log out. The nature of Linux is the establishment of security and permissions, so determine how much protection your system requires and design it accordingly.

Strictly control the access to your / root directory. Linux isn't like other software you use on a daily basis; it doesn't question your intentions. Telling the software to wipe out the / root directory results in the quick and efficient elimination of this directory. It doesn't expect you to make mistakes. It expects you to know what you are doing. You won't get multiple pop-up windows asking you to confirm your actions. For this reason, root access in Linux isn't for the novice or the squeamish. One mistake could mean reinstalling Linux and all the bits and pieces it once held.

Knowing the difference between sudo and su

Asterisk has two commands that look similar, sound similar, and function similarly, but are different: `sudo` and `su`.

The `sudo` command functions like an application. Executing a `sudo` command allows a specified user to run a program in Linux as a pseudoversion of a different user, generally a root user.

The `su` (switch user) command requests a new login and password from Linux, which allows you to change the user you're logged in as.

You can execute the `su` command as any of the following depending on what you're doing:

✔ This command defaults to the login for the *super user* that has access to all directories, including the root:

```
su
```

✔ This command prompts you for the root password:

```
su root
```

✔ This command prompts you to log in as the user named Stephen, which grants you all the privileges and access assigned to Stephen:

```
su stephen
```

In a nutshell, the `sudo` command allows you to execute a command as if you were another user (generally a root user) if authorized, while you're still logged in as yourself. The `su` command is simpler and allows you to simply log out as one user and log in as another. After you are logged in as the new user, Linux treats you as that user, with no connection to the last user you logged in as.

Many people like to use the `su` command because it prevents them from accidentally damaging vital aspects of the Linux. If you don't need to have root access for a given task, it is generally safer to remain logged in as a nonroot user.

Visiting the directory neighborhood

The root directory isn't the only directory housed within Linux. The Asterisk files you are installing and configuring aren't in the main root directory. Linux contains these directories:

- ✔ **/:** This is the main root directory, which contains all other directories and files.
- ✔ **/root:** This is the administration directory for the main root users directory.
- ✔ **/boot:** The files required to boot the server are in this directory. Your interaction with Asterisk never takes you into this file.

- ✔ /etc: This directory holds the configuration and startup files for applications on the Linux system. Installation of Asterisk occurs with files and subdirectories in this directory.

  You need root access to enter this directory and to modify the config files found within it.

- ✔ /bin: Binary and command files are stored here. Your Asterisk installation and programming don't take you to this directory.

- ✔ /sbin: This directory contains files and subdirectories for administration commands and super user (root user) information. The programs in this directory are file system programs and other lower-level and less-used programs. These files are only used for periodic administration.

- ✔ /usr: The programs and applications reside in this directory. This directory holds several Asterisk directories, including the following:

  - /usr/lib/asterisk, which contains the compiled library files
  - /usr/include/asterisk, which holds programming header libraries
  - /usr/sbin, where you find the executable programs for Asterisk

- ✔ /home: This directory contains the home directories for users. This is the first directory you arrive at when entering Linux as a user, and it's where you store all your personal Linux files.

- ✔ /var: The logs and system files are housed in this directory, including printers and mail.

- ✔ swap: This is a partition used by Linux and acts as a repository for files that are being transferred. It's like virtual memory for your computer, allowing a location to store data that's in transit between locations.

- ✔ /proc: The proc, or *processes,* directory doesn't contain files, but instead acts as a virtual file system containing information about files and the Linux system, such as the running processes and kernel parameters.

- ✔ /lost+found: The files and bits of data housed in this directory are the result of a Linux system crash or unexpected shut down of the system. Linux conducts a file system check when it re-boots from the outage, identifying any corrupted files and placing them in this directory. The files that end up in this directory need to be moved back to their original locations in Linux. This can be a tedious and time-consuming task, and in some cases may require a re-installation of Linux.

- ✔ /dev: Hardware devices are listed here, such as hard drives, floppy drives, and so on, along with their required drivers and supporting software.

Working with Files

Moving around within Linux is helpful, but when you land in a specific directory, you need to know how to select a file for editing and how to save it.

Linux filenames don't follow the Windows or DOS-based filename conventions. For example, a file in a Windows or DOS-based environment can end in `.exe` or `.bat`. The permissions in Linux determine whether a file is executable, not the file extension.

Editing files

Many programming and installation aspects of Asterisk require you to edit specific files. If you haven't worked in Linux before, this seems like a daunting task. Because the dialplan exists in the `extensions.conf` file, we show you how to modify it.

From the root or home directory, change to the Asterisk directory with the following command:

```
cd/etc/asterisk/
```

You move to the correct directory in which the `extensions.conf` file exists, but you still have to open the file for editing. Use the `vi` command to display the file for editing as follows:

```
vi extensions.conf
```

The `extensions.conf` file is now displayed on your screen. Pressing the `I` key allows you to insert text and add, change, or delete the contents of the file.

The `I` command is specific to the `vi` text editor. Other editors, such as `view`, may have different commands.

Saving files

After you modify a file, you have to save it. Press Esc to move from edit mode to command mode. After you're in command mode, type the following command to save the changes in the file:

```
:w
```

This command is also specific to `vi`, so if you're using another editor, you probably need to use a different command. All the changes you've made are saved, but the document is still available for changes. You can save the document several times just as you do with other editors, such as Microsoft Word. You'll remain in the `vi` command mode until you enter the I command to re-enter the document for editing. Find more information on `vi` in the user documentation at `http://vimdoc.sourceforge.net/htmldoc/usr_toc.html`.

Closing a file

You don't leave a document until you quit it. Press Esc and then type the following command to close the file and return to the `/etc/asterisk` directory:

```
:q
```

Configuration files like the `extensions.conf` file are in plain text and don't need to be recompiled. You'll know that the file you're working on must be recompiled if you open the file after editing and see garbled characters.

Backing up files

The *t*ape *ar*chive (*tar*) was originally created for tape backups. The intention was to gather files, but not compress them as a zip file does. The tar system creates one file out of many; that file is colloquially referred to as a *tarball*.

You use the advanced tarball technology when installing Asterisk and your packet-capture software, such as tcpdump or Wireshark.

You can also back up files — in this case, in a tarball named Elijah — with the following command:

```
tar -tvf Elijah.tar
```

To see the contents of a tarball, use this command:

```
tar -xvf Elijah.tar
```

This command does more than just allow you to see the filenames of the tarball. It both extracts the contents of the tarball and lists the filenames of the extracted files.

Getting Info from the man Files

Linux has manuals available for all the commands it supports. Retrieving the information in the manual is simple; just use the man command. The syntax for the man command is as follows:

```
man command_name
```

For example, use this command to make a new directory:

```
man mkdir
```

Use this command to remove a directory:

```
man rm
```

The command spills forth one page of information that you can advance by pressing the spacebar. After you read everything you want, press Q to quit the manual.

Before quitting the manual, you can export the information to a text file with the following command (replace *MyFileName* with the name you want to assign to the new text file):

```
man mkdir >MyFileName
```

After the information is in a text file, you can easily open, move, or print it.

Using the same filename for all manual files overwrites the previous contents. Use a different filename for each manual you export to prevent confusion.

Searching Files and Directories

If you ever find yourself lost in a sea of Linux, desperately searching for the file in your home directory, don't worry; you can use the grep command. This command comes in handy if you are one of those people

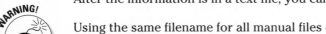

who continually lose car keys. It's the Linux version of a find command, allowing you to search the body of every document in the directory for a specific string of text.

Because you're searching for a file, you probably don't remember the specific wording within the text. This isn't a problem because the asterisk symbol (*) comes to your rescue. The * represents a *wildcard* in programming. The command looks like this:

```
grep mysearchterm *
```

For example, to search every line of code for every file within the directory that includes the name Brady, use this command:

```
grep brady *
```

To search for a term and then save the info to a file, add an output to the grep command, as follows:

```
grep mysearchterm * > myfilename
```

In the following example, we're searching every file in the directory for line items of text or code with the word brady, and saving the information to a BradyInfo file:

```
grep brady * > BradyInfo
```

If you're searching several times in succession and don't want to overwrite your file, use the following command:

```
grep mysearchterm * >> myfilename
```

Adding the second greater-than sign (>) in the code tells Linux to add the results to the bottom of the file, instead of overwriting it.

The grep command only searches the files in your directory. You can search every directory and subdirectory by executing the following command from the / root directory:

```
grep -R brady *
```

You can add logic to your grep command by adding a regular expression to your search. The egrep command, which uses the pipe symbol (|), representing either/or, is used in the following expression:

```
egrep mysearchterm|mysecondsearchterm *
```

For example, to find either jeanne or jonathan in the current directory, use the following command:

```
egrep jeanne|jonathan *
```

Index

• T •

BUSINESS, CAREERS & PERSONAL FINANCE

Fundraising FOR DUMMIES
0-7645-9847-3

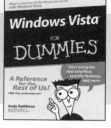

Investing FOR DUMMIES
0-7645-2431-3

Also available:
- Business Plans Kit For Dummies
  0-7645-9794-9
- Economics For Dummies
  0-7645-5726-2
- Grant Writing For Dummies
  0-7645-8416-2
- Home Buying For Dummies
  0-7645-5331-3
- Managing For Dummies
  0-7645-1771-6
- Marketing For Dummies
  0-7645-5600-2

- Personal Finance For Dummies
  0-7645-2590-5*
- Resumes For Dummies
  0-7645-5471-9
- Selling For Dummies
  0-7645-5363-1
- Six Sigma For Dummies
  0-7645-6798-5
- Small Business Kit For Dummies
  0-7645-5984-2
- Starting an eBay Business For Dummies
  0-7645-6924-4
- Your Dream Career For Dummies
  0-7645-9795-7

HOME & BUSINESS COMPUTER BASICS

Laptops FOR DUMMIES
0-470-05432-8

Windows Vista FOR DUMMIES
0-471-75421-8

Also available:
- Cleaning Windows Vista For Dummies
  0-471-78293-9
- Excel 2007 For Dummies
  0-470-03737-7
- Mac OS X Tiger For Dummies
  0-7645-7675-5
- MacBook For Dummies
  0-470-04859-X
- Macs For Dummies
  0-470-04849-2
- Office 2007 For Dummies
  0-470-00923-3

- Outlook 2007 For Dummies
  0-470-03830-6
- PCs For Dummies
  0-7645-8958-X
- Salesforce.com For Dummies
  0-470-04893-X
- Upgrading & Fixing Laptops For Dummies
  0-7645-8959-8
- Word 2007 For Dummies
  0-470-03658-3
- Quicken 2007 For Dummies
  0-470-04600-7

FOOD, HOME, GARDEN, HOBBIES, MUSIC & PETS

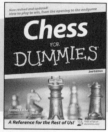

Chess FOR DUMMIES
0-7645-8404-9

Guitar FOR DUMMIES
0-7645-9904-6

Also available:
- Candy Making For Dummies
  0-7645-9734-5
- Card Games For Dummies
  0-7645-9910-0
- Crocheting For Dummies
  0-7645-4151-X
- Dog Training For Dummies
  0-7645-8418-9
- Healthy Carb Cookbook For Dummies
  0-7645-8476-6
- Home Maintenance For Dummies
  0-7645-5215-5

- Horses For Dummies
  0-7645-9797-3
- Jewelry Making & Beading For Dummies
  0-7645-2571-9
- Orchids For Dummies
  0-7645-6759-4
- Puppies For Dummies
  0-7645-5255-4
- Rock Guitar For Dummies
  0-7645-5356-9
- Sewing For Dummies
  0-7645-6847-7
- Singing For Dummies
  0-7645-2475-5

INTERNET & DIGITAL MEDIA

eBay FOR DUMMIES
0-470-04529-9

iPod & iTunes FOR DUMMIES
0-470-04894-8

Also available:
- Blogging For Dummies
  0-471-77084-1
- Digital Photography For Dummies
  0-7645-9802-3
- Digital Photography All-in-One Desk Reference For Dummies
  0-470-03743-X
- Digital SLR Cameras and Photography For Dummies
  0-7645-9803-1
- eBay Business All-in-One Desk Reference For Dummies
  0-7645-8438-3
- HDTV For Dummies
  0-470-09673-X

- Home Entertainment PCs For Dummies
  0-470-05523-5
- MySpace For Dummies
  0-470-09529-6
- Search Engine Optimization For Dummies
  0-471-97998-8
- Skype For Dummies
  0-470-04891-3
- The Internet For Dummies
  0-7645-8996-2
- Wiring Your Digital Home For Dummies
  0-471-91830-X

* Separate Canadian edition also available
† Separate U.K. edition also available

Available wherever books are sold. For more information or to order direct: U.S. customers visit www.dummies.com or call 1-877-762-2974.
U.K. customers visit www.wileyeurope.com or call 0800 243407. Canadian customers visit www.wiley.ca or call 1-800-567-4797.

WILEY

SPORTS, FITNESS, PARENTING, RELIGION & SPIRITUALITY

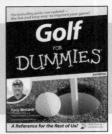

0-471-76871-5

0-7645-7841-3

Also available:

- Catholicism For Dummies
  0-7645-5391-7
- Exercise Balls For Dummies
  0-7645-5623-1
- Fitness For Dummies
  0-7645-7851-0
- Football For Dummies
  0-7645-3936-1
- Judaism For Dummies
  0-7645-5299-6
- Potty Training For Dummies
  0-7645-5417-4
- Buddhism For Dummies
  0-7645-5359-3

- Pregnancy For Dummies
  0-7645-4483-7 †
- Ten Minute Tone-Ups For Dummies
  0-7645-7207-5
- NASCAR For Dummies
  0-7645-7681-X
- Religion For Dummies
  0-7645-5264-3
- Soccer For Dummies
  0-7645-5229-5
- Women in the Bible For Dummies
  0-7645-8475-8

TRAVEL

0-7645-7749-2

0-7645-6945-7

Also available:

- Alaska For Dummies
  0-7645-7746-8
- Cruise Vacations For Dummies
  0-7645-6941-4
- England For Dummies
  0-7645-4276-1
- Europe For Dummies
  0-7645-7529-5
- Germany For Dummies
  0-7645-7823-5
- Hawaii For Dummies
  0-7645-7402-7

- Italy For Dummies
  0-7645-7386-1
- Las Vegas For Dummies
  0-7645-7382-9
- London For Dummies
  0-7645-4277-X
- Paris For Dummies
  0-7645-7630-5
- RV Vacations For Dummies
  0-7645-4442-X
- Walt Disney World & Orlando
  For Dummies
  0-7645-9660-8

GRAPHICS, DESIGN & WEB DEVELOPMENT

0-7645-8815-X

0-7645-9571-7

Also available:

- 3D Game Animation For Dummies
  0-7645-8789-7
- AutoCAD 2006 For Dummies
  0-7645-8925-3
- Building a Web Site For Dummies
  0-7645-7144-3
- Creating Web Pages For Dummies
  0-470-08030-2
- Creating Web Pages All-in-One Desk
  Reference For Dummies
  0-7645-4345-8
- Dreamweaver 8 For Dummies
  0-7645-9649-7

- InDesign CS2 For Dummies
  0-7645-9572-5
- Macromedia Flash 8 For Dummies
  0-7645-9691-8
- Photoshop CS2 and Digital
  Photography For Dummies
  0-7645-9580-6
- Photoshop Elements 4 For Dummies
  0-471-77483-9
- Syndicating Web Sites with RSS Feeds
  For Dummies
  0-7645-8848-6
- Yahoo! SiteBuilder For Dummies
  0-7645-9800-7

NETWORKING, SECURITY, PROGRAMMING & DATABASES

0-7645-7728-X

0-471-74940-0

Also available:

- Access 2007 For Dummies
  0-470-04612-0
- ASP.NET 2 For Dummies
  0-7645-7907-X
- C# 2005 For Dummies
  0-7645-9704-3
- Hacking For Dummies
  0-470-05235-X
- Hacking Wireless Networks
  For Dummies
  0-7645-9730-2
- Java For Dummies
  0-470-08716-1

- Microsoft SQL Server 2005 For Dummies
  0-7645-7755-7
- Networking All-in-One Desk Reference
  For Dummies
  0-7645-9939-9
- Preventing Identity Theft For Dummies
  0-7645-7336-5
- Telecom For Dummies
  0-471-77085-X
- Visual Studio 2005 All-in-One Desk
  Reference For Dummies
  0-7645-9775-2
- XML For Dummies
  0-7645-8845-1

HEALTH & SELF-HELP

0-7645-8450-2

0-7645-4149-8

Also available:

- Bipolar Disorder For Dummies
  0-7645-8451-0
- Chemotherapy and Radiation For Dummies
  0-7645-7832-4
- Controlling Cholesterol For Dummies
  0-7645-5440-9
- Diabetes For Dummies
  0-7645-6820-5* †
- Divorce For Dummies
  0-7645-8417-0 †

- Fibromyalgia For Dummies
  0-7645-5441-7
- Low-Calorie Dieting For Dummies
  0-7645-9905-4
- Meditation For Dummies
  0-471-77774-9
- Osteoporosis For Dummies
  0-7645-7621-6
- Overcoming Anxiety For Dummies
  0-7645-5447-6
- Reiki For Dummies
  0-7645-9907-0
- Stress Management For Dummies
  0-7645-5144-2

EDUCATION, HISTORY, REFERENCE & TEST PREPARATION

0-7645-8381-6

0-7645-9554-7

Also available:

- The ACT For Dummies
  0-7645-9652-7
- Algebra For Dummies
  0-7645-5325-9
- Algebra Workbook For Dummies
  0-7645-8467-7
- Astronomy For Dummies
  0-7645-8465-0
- Calculus For Dummies
  0-7645-2498-4
- Chemistry For Dummies
  0-7645-5430-1
- Forensics For Dummies
  0-7645-5580-4

- Freemasons For Dummies
  0-7645-9796-5
- French For Dummies
  0-7645-5193-0
- Geometry For Dummies
  0-7645-5324-0
- Organic Chemistry I For Dummies
  0-7645-6902-3
- The SAT I For Dummies
  0-7645-7193-1
- Spanish For Dummies
  0-7645-5194-9
- Statistics For Dummies
  0-7645-5423-9

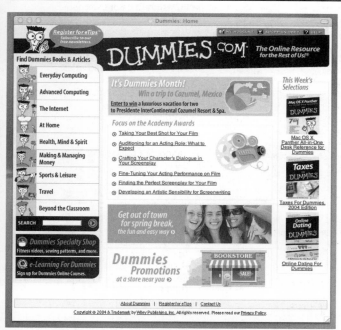

Get smart @ dummies.com®

- **Find a full list of Dummies titles**
- **Look into loads of FREE on-site articles**
- **Sign up for FREE eTips e-mailed to you weekly**
- **See what other products carry the Dummies name**
- **Shop directly from the Dummies bookstore**
- **Enter to win new prizes every month!**

* Separate Canadian edition also available
† Separate U.K. edition also available

Available wherever books are sold. For more information or to order direct: U.S. customers visit www.dummies.com or call 1-877-762-2974.
U.K. customers visit www.wileyeurope.com or call 0800 243407. Canadian customers visit www.wiley.ca or call 1-800-567-4797.